Effects of Inlet Geometry on

Hydraulic Performance of Box Culverts

November 15, 2007

Errata for Effects of Inlet Geometry on Hydraulic Performance of Box Culverts Publication No. (FHWA-HRT-06-138)

Dear Customer:

A number of technical corrections were made to this report after the report was originally published on the Web site. The following table shows the modifications that were made to this electronic report.

Page Number	Correction
Page 42	Table 1 replaced
Page 47 and 48	Table 2 replaced
Page 59	Table 3 replaced
Page 62	Table 4 replaced
Page 73	Table 7 replaced
Page 73	Table 8 replaced
Page 74	Table 9 replaced
Page 74	Table 10 replaced
Page 85	Table 11 replaced
Page 86	Table 12 replaced
Page 120 and 121	Table 17 replaced
Page 122 and 123	Table 18 replaced

FOREWORD

This report describes a laboratory study of culvert hydraulics done at the TFHRC hydraulics lab in partnership with the South Dakota DOT (SDDOT). The study focused on rectangular-shaped culverts with a number of inlet geometry conditions representing inlets that are currently available for highway culverts. Design coefficients are recommended for several inlet configurations that are not specifically covered in the Federal Highway Administration Hydraulic Design Series No. 5 (HDS-5). This report will be of interest to hydraulic engineers involved in culvert design and to researchers involved in developing improved culvert design guidelines. It is being published as a Web document only.

Gary Henderson
Director, Office of Infrastructure
Research and Development

1. Report No	2. Government Accession No.	3. Recipient's Catalog No.		
FHWA-HRT-06-138	N/A	N/A		
4. Title and Subtitle		5. Report Date		
Effects of Inlet Geometry on Hydraulic Performance of Box Culverts		December 2006		
		6. Performing Organization Code		
		N/A		
7. Authors(s)		8. Performing Organization Report No.		
J. Sterling Jones, Kornel Kerenyi, and Stuart Stein		N/A		
9. Performing Organization Name and Address		10. Work Unit No. (TRAIS)		
GKY and Associates, Inc. 5411-E. Backlick Road Springfield, VA 22151		N/A		
		11. Contract or Grant No.		
		DTFH61—04-C-00037		
		13. Type of Report and Period Covered		
		Final Lab Report 11-02 to 11-04		
12. Sponsoring Agency Name and Address		14. Sponsoring Agency Code		
Office of Infrastructure R&D Federal Highway Administration 6300 Georgetown Pike McLean, VA 22101	South Dakota DOT Office of Research 700 E. Broadway Avenue Pierre, SD 57501	SD Project SD2002-04 FHWA Task Order 3		

15. Supplementary Notes

Contracting Officer's Technical Representative: J. Sterling Jones
South Dakota Technical Panel: Mark Clausen, Noel Clocksin, Kevin Goeden, Cory Haeder, Terry Jorgensen, Paul Oien, Daris Ormesher, Rich Phillips, Steve Wagner
Brad Newlin designed models and developed preliminary analysis algorithms. Holger Dauster and Amon Tarakemeh provided invaluable assistance with instrumentation, data collection, and analysis.
Donna and Dave Pearson provided senior editing and EDP services.

16. Abstract

Each year, the South Dakota Department of Transportation (SDDOT) designs and builds many cast-in-place (CIP), or field cast, and precast box culvert structures that allow drainage to pass under roadways. The CIP boxes typically have 30-degree-flared wingwalls, and the precast have straight wingwalls with 10.16-centimeter (cm) (4-inch) bevels on the inside edges of the wingwalls and top slab. Previous research conducted on a limited number of single barrel box culverts indicated that further research was necessary to determine (1) the effects of multiple barrel structures, (2) loss coefficients of unsubmerged outlets, and (3) the effects of 30.48-cm (12-inch) corner fillets versus 15.24-cm (6-inch) corner fillets. In order to optimize the design of both types of box culverts, it was also necessary to determine the effects of span-to-rise ratios, skewed end conditions, and optimum edge conditions on typical box culvert installations

17. Key Words	18. Distribution Statement			
Inlet geometry, box culverts	No restrictions. This document is available to the public through the National Technical Information Service; Springfield, VA 22161			
19. Security Classif. (of this report)	20. Security Classif. (of this page)		21. No. of Pages	22. Price
Unclassified	Unclassified		158	

Form DOT F 1700.7 (8-72)   Reproduction of completed page authorized (art. 5/94)

SI* (MODERN METRIC) CONVERSION FACTORS

APPROXIMATE CONVERSIONS TO SI UNITS

Symbol	When You Know	Multiply By	To Find	Symbol
LENGTH				
in	inches	25.4	millimeters	mm
ft	feet	0.305	meters	m
yd	yards	0.914	meters	m
mi	miles	1.61	kilometers	km
AREA				
in^2	square inches	645.2	square millimeters	mm^2
ft^2	square feet	0.093	square meters	m^2
yd^2	square yard	0.836	square meters	m^2
ac	acres	0.405	hectares	ha
mi^2	square miles	2.59	square kilometers	km^2
VOLUME				
fl oz	fluid ounces	29.57	milliliters	mL
gal	gallons	3.785	liters	L
ft^3	cubic feet	0.028	cubic meters	m^3
yd^3	cubic yards	0.765	cubic meters	m^3
NOTE: volumes greater than 1000 L shall be shown in m^3				
MASS				
oz	ounces	28.35	grams	g
lb	pounds	0.454	kilograms	kg
T	short tons (2000 lb)	0.907	megagrams (or "metric ton")	Mg (or "t")
TEMPERATURE (exact degrees)				
°F	Fahrenheit	5 (F-32)/9 or (F-32)/1.8	Celsius	°C
ILLUMINATION				
fc	foot-candles	10.76	lux	lx
fl	foot-Lamberts	3.426	candela/m^2	cd/m^2
FORCE and PRESSURE or STRESS				
lbf	poundforce	4.45	newtons	N
lbf/in^2	poundforce per square inch	6.89	kilopascals	kPa

APPROXIMATE CONVERSIONS FROM SI UNITS

Symbol	When You Know	Multiply By	To Find	Symbol
LENGTH				
mm	millimeters	0.039	inches	in
m	meters	3.28	feet	ft
m	meters	1.09	yards	yd
km	kilometers	0.621	miles	mi
AREA				
mm^2	square millimeters	0.0016	square inches	in^2
m^2	square meters	10.764	square feet	ft^2
m^2	square meters	1.195	square yards	yd^2
ha	hectares	2.47	acres	ac
km^2	square kilometers	0.386	square miles	mi^2
VOLUME				
mL	milliliters	0.034	fluid ounces	fl oz
L	liters	0.264	gallons	gal
m^3	cubic meters	35.314	cubic feet	ft^3
m^3	cubic meters	1.307	cubic yards	yd^3
MASS				
g	grams	0.035	ounces	oz
kg	kilograms	2.202	pounds	lb
Mg (or "t")	megagrams (or "metric ton")	1.103	short tons (2000 lb)	T
TEMPERATURE (exact degrees)				
°C	Celsius	1.8C+32	Fahrenheit	°F
ILLUMINATION				
lx	lux	0.0929	foot-candles	fc
cd/m^2	candela/m^2	0.2919	foot-Lamberts	fl
FORCE and PRESSURE or STRESS				
N	newtons	0.225	poundforce	lbf
kPa	kilopascals	0.145	poundforce per square inch	lbf/in^2

*SI is the symbol for the International System of Units. Appropriate rounding should be made to comply with Section 4 of ASTM E380.
(Revised March 2003)

TABLE OF CONTENTS

LIST OF FIGURES

v

LIST OF TABLES

LIST OF ACRONYMS AND ABBREVIATIONS

2-D	two-dimensional
ccd	charge coupled device
CIP	cast-in-place
DOT	department of transportation
EGL	energy grade line
FC	field cast
FHWA	Federal Highway Administration
HDS	Hydraulic Design Series
HECRAS	Hydrologic Engineering Center River Analysis System
HGL	hydraulic grade line
HW/D	headwater depth (ratio)
LDA	laser doppler anemometry
NCHRP	National Cooperative Highway Research Program
OPM	optical pressure measurement
PC	precast
PIV	particle image velocimetry
PVC	polyvinyl chloride
SI	International System of Units (the Metric System)
SDDOT	South Dakota Department of Transportation
TFHRC	Turner-Fairbank Highway Research Center
VI	virtual instruments
WW	wingwall

LIST OF SYMBOLS
(A comprehensive list can be found in *Hydraulic Design of Highway Culverts* (HDS-5)[1])

A	full cross-sectional area of culvert barrel
c	coefficient for submerged inlet control equation
D	interior height of the culvert barrel
EGL	energy grade line (sometimes E.G.L.)
h	height of hydraulic grade line above centerline of orifice
H_c	specific head at critical depth ($d_c + V_c^2/2g$)
H_e	entrance head loss
H_f	friction head loss in culvert barrel
H_L	total energy loss
H_{Le}, H_{Lc}	inlet loss (also the inlet head loss or the contraction loss)
H_{Lf}, H_f	friction loss
H_{Lo}	exit loss
H_o	exit head loss
HGL	hydraulic grade line (sometimes H.G.L.)
HW	headwater; depth from inlet invert to upstream total energy grade line
HW_i	headwater depth above inlet control section invert
HW_o	headwater depth above culvert outlet invert

HW/D	headwater depth ratio
g	acceleration due to gravity
K	coefficient for unsubmerged inlet control equation
K_e	coefficient for outlet control entrance loss
K_o	exit loss coefficient usually assumed to be 1.0 for design purposes
K_u	1.811 for SI; 1.0 for the English system
M	exponent in unsubmerged inlet control equation
Q	discharge
S	culvert barrel slope
TW	tailwater; depth of water measured from culvert outlet invert
V	mean velocity of flow
V_d	downstream velocity
V_u	approach (upstream) velocity
y	depth of flow
Y	additive term in submerged inlet control equation

ABBREVIATED GLOSSARY

Head	a measure of water.
Headwater	depth of the upstream water measured from the invert at the culvert entrance.
Head loss	a measure of the reduction of the total head of a fluid as it moves through a fluid system.
Invert	the inside bottom elevation of a closed conduit, including a culvert.
Reynolds number	the ratio of inertial forces to viscous forces.
Tailwater	depth of water downstream of the culvert measured from the outlet invert.
Vena contracta	point of minimum area of a flow.

CHAPTER 1. INTRODUCTION

The South Dakota Department of Transportation (SDDOT) and the Federal Highway Administration (FHWA) collaborated on a research study conducted at the Turner-Fairbank Highway Research Center (TFHRC) hydraulics laboratory to determine the effects of a number of inlet geometry choices on culvert hydraulic efficiency. This study is a response to the large number of culverts that are installed in the United States and the fact that most of the current guidelines on culvert hydraulics are based on research completed more than 20 years ago. A conservative estimate indicates that there are more than 3.66 million linear meters (12 million linear feet) of culverts installed in the United States every year. The most widely recognized manual on culvert hydraulics is the FHWA Hydraulic Design Series No. 5 (HDS-5), *Hydraulic Design of Highway Culverts,*[1] published in 1985 but based on research conducted in the 1960s and 1970s. Most State DOT engineers use the FHWA HY-8 computer program[2] or similar programs based on HDS-5 for hydraulic evaluation and design of highway culverts. It is important to implement new technology in these programs to benefit practitioners in the State DOTs. Results from this study are presented in a format that is similar to HDS-5 to facilitate implementation in these programs.

PROBLEM STATEMENT

Each year, SDDOT designs and builds many cast-in-place (CIP) and precast box culvert structures that allow drainage to pass under roadways. The CIP boxes typically have 30-degree flared wingwalls and the precast have straight wingwalls with a 10.2-centimeter (cm) (4-inch) bevel on the inside edges of the wingwalls and top slab. An analysis of previous research, that research being described in *South Dakota Culvert Inlet Design Coefficients,*[3] conducted on a limited number of single barrel box culverts, indicated that further research was necessary to determine (1) the effects of multiple barrel structures, (2) loss coefficients of unsubmerged inlets, and (3) the effect of 30.5-cm (12-inch) corner fillets versus 15.2-cm (6-inch) corner fillets. In order to optimize the designs of both types of box culverts, the effects of the span-to-rise ratio, skewed end condition, and optimum edge condition should also be determined.

A major problem with the current analysis programs for sizing box culvert structures (HY-8 and others) is that they do not analyze multiple barrel box culverts correctly. These programs model multiple barrel structures as though each barrel is a separate single box with its own wingwalls. Multiple barrel structures, however, share a single set of wingwalls. Most CIP box culverts fall in this category of multiple barrel structures with a single set of wingwalls.

In the case of a wingwall configuration and a single barrel, the wingwalls conduct the flow directly into the barrel, reducing the contraction losses at the entrance. For the same configuration with multiple barrels, there is minimal contraction loss for interior barrels so losses are much lower. In other words, multiple barrels should perform better than a single barrel multiplied by the number of barrels.

OBJECTIVES

The objectives of this study were to:

- Determine optimum edge conditions for wingwalls.
- Determine the effects of inlet geometry on flow capacity of single and multiple barrel culverts with optimized edge treatment of wingwalls.
- Determine effects of the span-to-rise ratio on flow capacity with various inlet geometries.
- Determine the effects of skew on flow capacity of box culverts.

PROCEDURES AND FACILITIES

An SDDOT technical review panel worked with the FHWA research team to develop a test matrix that included six edge conditions and 32 inlet configurations for rectangular box culverts to be tested at two slopes, two tailwater conditions, and various discharge intensities. A total of approximately 680 tests were conducted in a special culvert test facility built for the study. A 1:12 model scale was selected for the test facility and very precise Plexiglas™ models were fabricated to isolate various features of inlet geometry. The inlet models were fabricated with clip-on components so that it was relatively easy to mix and match components to isolate any feature without switching whole models.

The experimental setup included three subsystems: a culvert barrel, a headbox, and a tailbox. The headbox and tailbox had Plexiglas walls, which were supported by a metal frame. Figure 1 shows the headbox under construction.

Figure 1. Photo. Culvert headbox under construction.

The headbox could be modified to vary the width of the approach flow. The height of the tailbox was adjustable to analyze different barrel slopes. The culvert barrel was made from a Plexiglas pipe. Ceramic class pressure sensors (pressure range: 0–10 kilopascal (kPa)(0–1.45 poundforce per square inch (lbf/inch^2))) were mounted in the centerline on the bottom of the experimental setup (figure 2) to measure instantaneous hydraulic grade lines. Taking time averages led to more precise loss coefficient computations. The discharge was provided by a 0.140 cubic meter per second (m³/s) (5 cubic feet per second (ft³/s), computer controlled pump. Flow depths and mean velocities were computed from pressure sensor measurements in the culvert barrel where flow was parallel to the invert. In the highly turbulent region in the vicinity of the culvert inlet and in the headbox where the transverse flow distribution was extreme, particle image velocimetry (PIV) and/or velocity probes augmented these measurements.

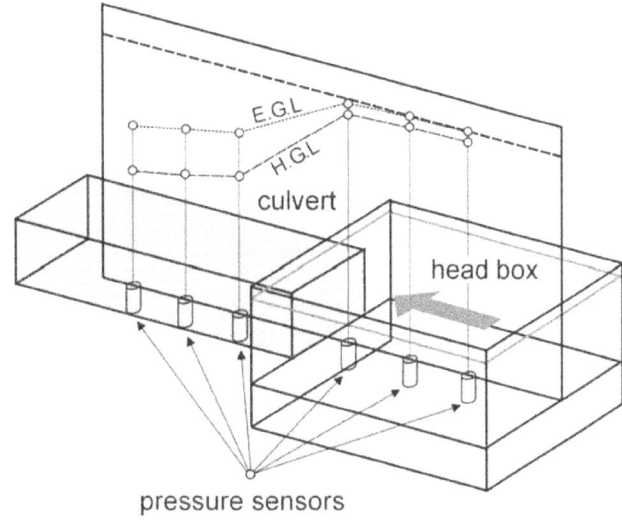

Figure 2. Diagram. Arrangement of the ceramic class pressure sensors.

PIV measures instantaneous velocity flow fields. It uses a focused light source, a high-resolution digital camera, and sophisticated computer logic to trace particle movements. The technique can accurately measure velocity in complex situations such as flows into culverts.

CHAPTER 2. LITERATURE REVIEW

The most comprehensive publication available in the literature is the FHWA HDS-5, *Hydraulic Design of Highway Culverts*,[1] a synthesis of culvert research that includes the classic studies done for the Bureau of Public Roads by the National Bureau of Standards during the 1950s and 1960s. (See references 4, 5, 6, 7, 8, 9, and 10.) HDS-5 features sections on design considerations, conventional culvert design, tapered inlets for various types of culverts, storage routing, and special considerations. Appendixes include design methods and equations, barrel resistance, design optimization using performance curves, and design charts, tables, and forms.

HDS-5 defines culvert hydraulics in terms of inlet and outlet control depending on the variables that influence the head required to push flow through the barrel. Inlet control occurs for steep culverts flowing free surface where flow goes through critical depth near the inlet. Flow in the culvert barrel below the critical depth section is supercritical flow that does not propagate downstream surface disturbances upstream. The only variables that affect the headwater are the discharge intensity and the geometry of the inlet. Outlet control occurs for mild slope culverts where free surface flow is subcritical and for any slope when the barrel is completely submerged. In these cases, the tailwater, which is typically known, is the control, and the headwater is affected by tailwater depth, outlet loss, friction loss, elevation difference, and the entrance loss, which is a function of discharge intensity and inlet geometry.

Outlet control is the more general case where the entrance loss is just one component that affects the headwater and usually is not the dominant component compared to the tailwater elevation and the friction in the barrel. The entrance loss is assumed to be a fraction of the velocity head in the barrel and is expressed as a coefficient times the velocity head. HDS-5 lists the entrance-loss coefficients as a single, constant value for each inlet shape. There is no distinction between high flows and low flows, but HDS-5 is a hydraulic design manual; so it is reasonable to expect coefficients to be more related to high flows. Furthermore, it is reasonable to expect some variation in this coefficient at low flows because the effective inlet shape changes when only a portion of it is in the flow zone.

Inlet control is the special case where the inlet geometry and corresponding entrance loss is the dominant component that affects the headwater. Regression equations have been developed for each inlet shape to express headwater as a function of discharge intensity directly or to compute a loss component that can be added to the critical head-to-yield headwater. These regression equations apply for a range of discharge intensities that include low flows. HDS-5 lists the regression coefficients for predetermined equation forms for each inlet shape.

During the late 1980s, there was considerable interest in the hydraulics of long-span culverts, which were frequently proposed as low-cost alternatives to short bridges. Laboratory experiments at the FHWA Hydraulics Laboratory were conducted to investigate effects of some of the characteristic features of long-span culverts, namely the culvert shape, the span-to-rise ratio, and the contraction ratio.[11] Experiments were conducted in a 1.8-meter (m)- (6-foot (ft)-) wide by 21.4-m- (70-ft-) long tilting flume set at a slope expected to generate inlet control. Culvert shapes included circular (which was used as a benchmark), semicircular, high-profile arches, and metal box geometries commonly used for long-span installations. Inlet geometry for all shapes was a thin, projecting edge with no flared wingwalls. Culvert shape seemed to have very little effect at the higher discharges for submerged flow, but the high-profile arch shape appeared to have lower relative entrance losses at the lower discharges for unsubmerged flow. The study found no logical explanation for the apparent advantage for the high-profile arch at low flows.

The span-to-rise ratio was varied by testing three metal box culvert geometries referred to as a high box, a mid box, and a low box with span-to-rise ratios of 2.0, 3.25, and 4.5, respectively. The span was held constant (at 50.8 cm (20 inches)) while the rise was varied. The shapes varied slightly because the metal boxes were not actually rectangles; they had rounded corners and resembled arches more than they did rectangles. The general trend was the higher the span-to-rise ratio, the lower the efficiency. In other words, for a thin edge projecting inlet where there was no bevel to streamline the flow over the top edge, increasing the span-to-rise ratio actually increased the headwater required to convey a given discharge intensity through the inlet.

The contraction ratio—approach channel width divided by culvert width—varied from 6.0 to 1.5. It appeared that the lower the contraction ratio, the higher the efficiency. But the primary conclusion drawn from this part of the study was that the headwater in HDS-5 was the specific energy head and not just the hydraulic grade line depth as is often presumed. To make the data agree with the performance curves shown in HDS-5 for the benchmark shape, it was necessary to include the approach-flow velocity head in the headwater computations. Typically, long-span culverts are nearly the full width of the approach channel, the contraction ratios are small, and the approach flow velocity is almost as high as the velocity in the culvert. This particular FHWA study was conducted to gain insight about the hydraulics of long-span culverts, but the results were never published.

A study conducted at the FHWA Hydraulics Laboratory for SD DOT compared hydraulic performance of precast inlet configurations to traditional 30-degree-flared wingwall inlets for box culverts.[3] Six culvert models constructed of plywood were tested for both inlet and outlet control. Water depths were measured through ports in the flooring via Tygon® tubing connected to a pressure transducer. Box culverts with single 1.8- by 1.8-m (6- by 6-ft), 2.4- by 2.4-m (8- by 8-ft), 2.7- by 2.7-m (9- by 9-ft), and 3.7- by 3.7-m (12- by 12-ft) barrels with 30-degree wingwalls were modeled in this study. Model scales of 1:10.67, 1:15, and 1:16 were selected to use stock thickness materials to simulate culvert wall thickness and wingwall thickness. Two slopes—3 percent and 1.75 percent—were used in the experiments. Effects of wingwall miters (to the embankment slope), straight-cut bevels, culvert barrel slopes, wingwall flare, and parapets were compared.

Inlet-control design coefficients were developed by regressing experimental data using the inlet-control design equations found in HDS-5. A benchmark culvert model was fabricated and tested to compare with scale 3 of the HDS-5 chart 8 as a check on experimental procedures. Inlet-control coefficients were derived for unsubmerged and submerged conditions for each culvert model, and the outlet control entrance-loss coefficient, K_e, was computed for each culvert model.

For inlet control, the design coefficients for the benchmark model of the HDS-5 chart 8, scale 3, did not match the values tabulated in HDS-5 very well; however, the outlet control coefficient, K_e, experimental value of 0.68 was a close match to the tabulated value of 0.7. For unsubmerged conditions, the miter slope, span-to-rise ratio, and culvert barrel slope appeared to have insignificant effect on the design coefficients. For submerged conditions, the 3:1 miter was slightly more efficient than a 2:1 miter. In contrast to the observation noted for the long-span culvert study, the higher span-to-rise ratios improved culvert performance (reduced headwater for a given discharge intensity), but these models did not have the thin edge projecting inlet geometry. Parapets used to retain fill over the top plate appeared to improve rather than hinder culvert performance.

Overall, the precast inlets with beveled edges were slightly better than the typical field cast inlets without beveled edges but were not as good as the 30-degree-flared wingwall inlet. No attempt, however, was made in the study to optimize the bevels. A number of general trends were noted, but there were no recommendations about how to modify FHWA manuals or computer programs to implement results from the study.

A study for the Iowa DOT by Graziano, et al., also conducted at the FHWA Hydraulics Laboratory, investigated the hydraulic performance of special Iowa DOT slope-tapered pipe culverts.[12] The culverts consisted of off-the-shelf components including precast end sections, one-eighth bends, and pipe reducers that are readily available from pipe suppliers. The goals of the study were to derive design coefficients for a slope-tapered inlet for circular culverts and to investigate the sensitivity of performance to reducer length and taper ratio. The performance of the precast end section, which was a flared-transition section that conformed to a 3:1 embankment slope, was compared to the performance of the HDS-5 chart 1, scale 1, culvert, which is a circular concrete culvert with a headwall and square edges.

Model scale ratios of 1:6.783 and 1:4.174 were used for the study. The headbox and tailbox were plywood versions of the culvert test facility that is currently in the laboratory. Hydraulic depths were measured by a single-pressure transducer connected through a switching block to pressure ports located along the culvert invert, in the headbox and in the tailbox. An adjustable tailgate was used to submerge the culvert to develop outlet control for a steep culvert.

For inlet control, the precast end section by itself without the other components for the slope-tapered inlet performed almost the same as the HDS-5 chart 1, scale 1, inlet. When the precast end section was combined with the reducers and bends to make the Iowa slope-tapered unit, hydraulic performance improved substantially. Performance was not sensitive to the taper ratio or whether one, two, or three reducers were used to transition the taper.

For outlet control, the tabulated K_e value for the HDS-5 chart 1, scale 1, inlet was 0.50, compared to a K_e of 0.35 for the precast end section and a K_e of 0.20 for the Iowa slope-tapered inlet. For inlet control, the design coefficients for the precast end section were K is 0.51 and M is 0.55 for the unsubmerged form 2 equation and c is 0.021 and Y is 0.823 for the submerged flow equation. K is the coefficient for the unsubmerged inlet control equation; M is the exponent for the unsubmerged inlet control equation; c is the coefficient for the submerged inlet control equation; and Y is an additive term in the submerged inlet control equation. The corresponding coefficients for the Iowa slope-tapered inlet were: K = 0.477; M = 0.533; c = 0.025; and Y = 0.659.

GKY & Associates, Inc. consolidated design coefficients, including the fifth-order polynomials that were used to code computer programs such as HY-8.[13] Derivations for the various equations cited in HDS-5, a comprehensive set of design coefficients, and nomographs or performance curves for all of the inlets covered by HDS-5 plus several that were studied later were included in this report.

McEnroe and Johnson tested shop fabricated metal and precast concrete open-flared end sections that are commonly available from pipe suppliers.[14] They also studied the effects of flow bars and debris blockage on the hydraulic performance of the pipes. They noted that HDS-5 provides little information on the hydraulic characteristics of these common end sections other than from limited hydraulic tests hydraulically equivalent in operation to a headwall in both inlet and outlet control. Their experiments with two pipe sizes resulted in outlet control K_e values ranging from 0.24 to 0.31. Both the metal and concrete end sections had the larger value for the smaller pipe size and the lower value for the larger pipe size.

The precast concrete open flare end section tested by McEnroe and Johnson was the same as the end section tested by Graziano, et al., for the Iowa DOT[12] and is illustrated in figure 3. Graziano recommended a K_e value of 0.35, which is comparable to the 0.31 value found by McEnroe and Johnson for that end section.

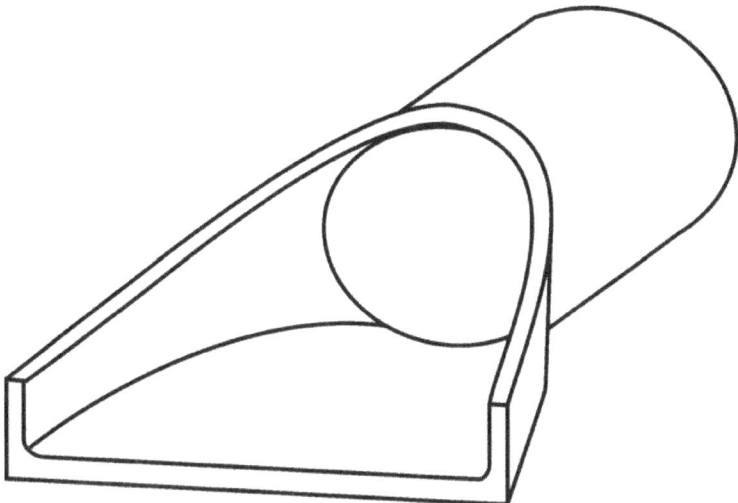

Figure 3. Sketch. Precast flared end section tested by Graziano and by McEnroe.

McEnroe and Johnson did not follow the HDS-5 pattern for presenting the inlet-control results; rather, they provided three dimensionless component equations for each inlet. The first component equation essentially represents the unsubmerged condition. The second one represents the transition zone, and the third one represents the submerged condition. Figures 4, 5, and 6 are the component equations for the prefabricated metal end section. In figures 4 through 10, the terms are as follows: HW is the headwater, or the depth from the inlet invert to the upstream total energy grade line; D is the interior height of the culvert barrel; Q is the discharge; and g is the acceleration due to gravity.

$$\frac{HW}{D} = 1.60 \left(\frac{Q}{\sqrt{g\,D^5}} \right)^{0.60} \quad for\ 0 \le \frac{Q}{\sqrt{g\,D^5}} \le 0.41$$

Figure 4. Equation. HW/D, prefabricated metal end section, unsubmerged condition.

$$\frac{HW}{D} = 2.23 \left(\frac{Q}{\sqrt{g\,D^5}} \right) + 0.023 \quad for\ 0.41 \le \frac{Q}{\sqrt{g\,D^5}} \le 0.62$$

Figure 5. Equation. HW/D, prefabricated metal end section, transition zone.

$$\frac{HW}{D} = 1.289 - 1.61 \left(\frac{Q}{\sqrt{g\,D^5}} \right) + 2.90 \left(\frac{Q}{\sqrt{g\,D^5}} \right)^2 \quad for\ 0.62 \le \frac{Q}{\sqrt{g\,D^5}} \le 1.20$$

Figure 6. Equation. HW/D, prefabricated metal end section, submerged condition.

Figures 7, 8, and 9 are the component equations for the precast concrete end section.

$$\frac{HW}{D} = 1.53 \left(\frac{Q}{\sqrt{g\,D^5}} \right)^{0.55} \quad for\ 0 \le \frac{Q}{\sqrt{g\,D^5}} \le 0.42$$

Figure 7. Equation. HW/D, precast concrete end section, unsubmerged condition.

$$\frac{HW}{D} = 2.13 \left(\frac{Q}{\sqrt{g\,D^5}} \right) + 0.055 \quad for\ 0.42 \le \frac{Q}{\sqrt{g\,D^5}} \le 0.68$$

Figure 8. Equation. HW/D, precast concrete end section, transition zone.

$$\frac{HW}{D} = 1.367 - 1.50 \left(\frac{Q}{\sqrt{g\,D^5}} \right) + 2.50 \left(\frac{Q}{\sqrt{g\,D^5}} \right)^2 \quad for\ 0.68 \le \frac{Q}{\sqrt{g\,D^5}} \le 1.30$$

Figure 9. Equation. HW/D, precast concrete end section, submerged condition.

The first equation in each set, figures 4 and 7, can readily be converted to the HDS-5 format to compare with Graziano's results. Figure 10 is the converted first component for the precast concrete end section.

$$\frac{HW}{D} = K_u\ 0.515 \left(\frac{Q}{A\,D^{1/2}} \right)^{0.55} \quad for\ 0 \le \frac{Q}{A\,D^{1/2}} \le 3.04$$

Figure 10. Equation. Figure 8 in HDS-5 format.

Where: K_u = 1.0 for English units ; K_u = 1.38 for SI units; K_u is a coefficient for the unsubmerged inlet control equation.

Figure 10 is the form 2 equation for unsubmerged inlet control flow, which means McEnroe and Johnson obtained values for K and M of 0.51 and 0.55, respectively. These were the exact values obtained by Graziano in a completely independent study.

Umbrell, et al., did a site-specific model study of a culvert installation consisting of a larger concrete culvert in series with a smaller culvert.[15] The inlet was a 30-degree flared wingwall, which was tested with both culvert diameters. The K_e values ranged from 0.12 to 0.24, but the higher values tended to be for the larger pipe size and the lower discharges. Researchers attributed the differences to experimental scatter and proposed an average value of 0.14.

A current National Cooperative Highway Research Program (NCHRP) study at Utah State University is investigating the effects of culvert geometry on hydraulic performance for circular culverts with buried inverts, composite roughness, and other measures used to

10

promote fish passage through culverts. Entrance loss coefficients at low flows are of interest because fish passage criteria are based on average seasonal flows, not the higher flows used for culvert design.

Tullis observed that the entrance loss coefficients for outlet control were not constant values for an inlet geometry as normally presented in the literature.[16] There was considerable variation in the K_e values, especially for low flows. Researchers hypothesized that K_e might be a function of the Reynolds number as illustrated in the conceptual sketch shown in figure 11. The implication of that observation is somewhat distressing because it would tend to complicate an otherwise simple computation. The entrance loss coefficient seems to be higher at the lower Reynolds numbers, which means the larger culverts would tend to have lower loss coefficients. This hypothesis might help explain why McEnroe and Johnson and Umbrell, et al., found smaller K_e's for the larger models used in the tests.[14,15]

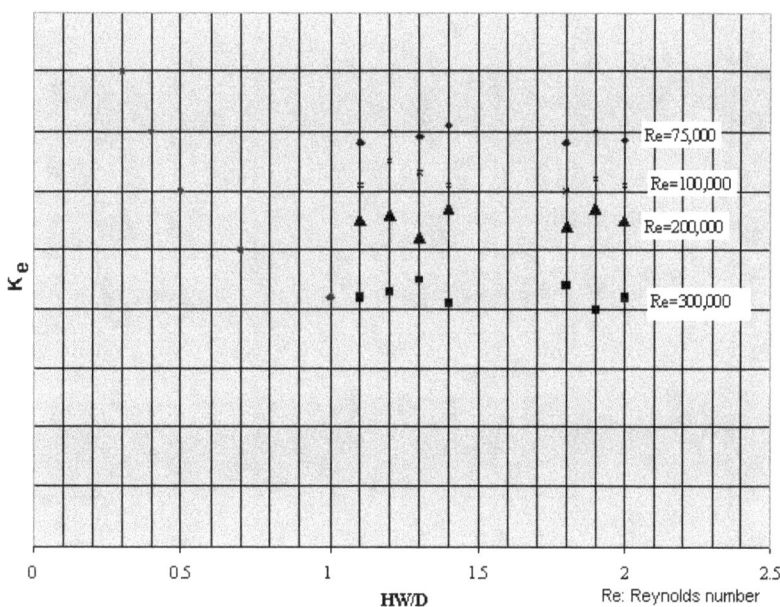

Note: The graph is hypothetical and meant to show the general relationship between K_e values and the Reynolds number. Specific K_e values are not given.

Figure 11. Sketch. Relationship of entrance loss coefficient to Reynolds number.

CHAPTER 3. THEORY AND DESIGN CALCULATIONS FOR INLET AND OUTLET CONTROL

The theory outlined in HDS-5, as described in this chapter, is the basis for analyzing data from this study.

INLET CONTROL HYDRAULICS OF CULVERTS

Empirical regression coefficients developed by fitting experimental data to semitheoretical equations are listed in HDS-5 for each inlet shape to relate headwater to discharge intensity for a range of flows.

Analysis of inlet-control data depends upon whether or not the inlet is submerged. If the inlet is not submerged, headwater can be expressed as a simple energy balance between the critical section and the upstream section, which are almost adjacent to one another. Alternatively, headwater can be analyzed as flow over a weir. Weir flow can be expressed as $Q = f(HW)^M$ or $HW = f(Q)^M$, where a typical value of M is 2/3 or 0.667. If the inlet is submerged, flow through the inlet is conceptually like flow through an orifice, albeit an irregular shaped orifice. Orifice flow can be expressed as $Q = cA(2gh)^{0.5}$ or $h = c(Q/A)^2$ where h is a depth measured from the center of the orifice. Since headwater depths are usually measured from the invert of a culvert, a more general expression for the orifice flow is $HW = c(Q/A)^2 + Y$, where Y is an offset distance.

Figure 12 illustrates an unsubmerged inlet control condition. Experimental data from this type of condition have been analyzed two ways. First, a simple energy balance can be made between sections 1 and 2 illustrated in figure 12. The critical specific energy H_c can be computed for any culvert shape, although it is tedious computation for some shapes. Data is regressed to determine regression coefficients K and M for computing the entrance loss, H_e. HDS-5 describes this analysis as the form 1 analysis for unsubmerged inlet control flow (figure 13). Most of the circular culvert inlets listed in HDS-5 are based on the form 1 analysis. The other way to analyze this type of data is to treat it as flow over a weir as illustrated in the form 2 equation (figure 14). Then the regression coefficients essentially absorb the critical specific energy control. All of the rectangular box culverts in HDS-5 are based on the form 2 equation, and in all cases, the exponent M was 0.667. Circular culverts and elliptical culverts with tapered throats are also based on the form 2 equation, but the exponent M varies a little for these inlets.

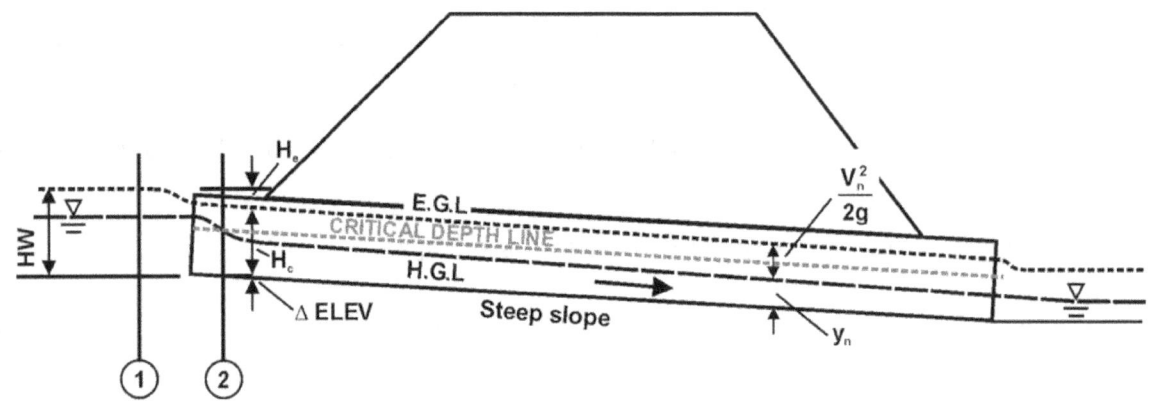

Figure 12. Diagram. Typical inlet control flow condition.

$$\frac{HW_i}{D} = \frac{H_c}{D} + K\left[\frac{K_u Q}{AD^{0.5}}\right]^M + 0.7S$$

Figure 13. Equation. Unsubmerged form 1, inlet control.

$$\frac{HW_i}{D} = K\left[\frac{K_u Q}{AD^{0.5}}\right]^M$$

Figure 14. Equation. Unsubmerged form 2, inlet control.

For inlet control under submerged conditions, regression constants c and Y are determined using the equation in figure 15.

$$\frac{HW_i}{D} = c\left[\frac{K_u Q}{AD^{0.5}}\right]^2 + Y + 0.7S$$

Figure 15. Equation. Submerged form, inlet control.

Where for figures 13, 14, and 15:

HW_i	is headwater depth above inlet control section invert.
D	is interior height of culvert barrel.
H_c	is specific head at critical depth ($d_c + V_c^2/2g$).
Q	is discharge.
A	is full cross-sectional area of culvert barrel.
S	is culvert barrel slope.
K, M, c, Y	are regression constants.
K_u	is 1.811 for SI and 1.0 for English units.

OUTLET CONTROL HYDRAULICS OF CULVERTS

Outlet control is the general case that covers everything except the special case of free surface flow passing through critical flow near the inlet. Outlet control flow occurs anytime the barrel flows full throughout, as illustrated in figure 16, regardless of the slope of the culvert.

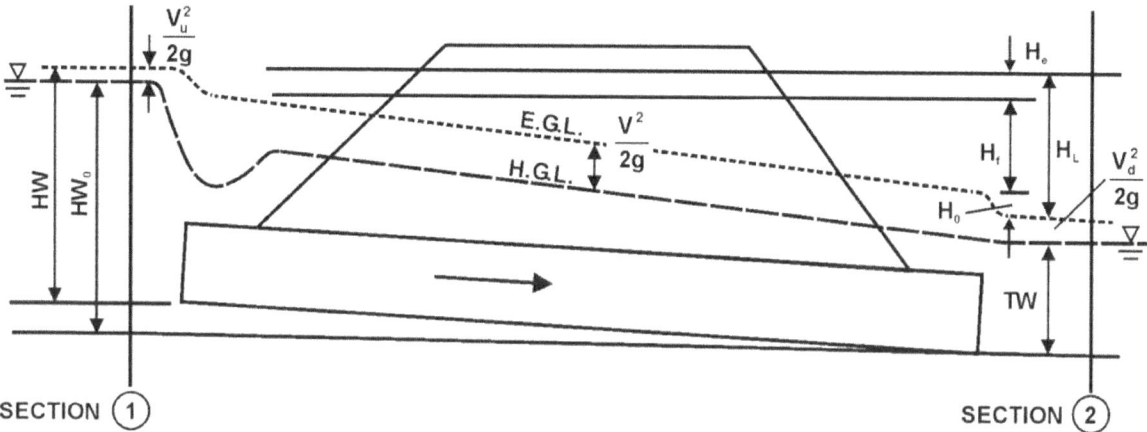

Figure 16. Diagram. Outlet control for full flow condition.

If the culvert is installed on a mild slope (0.7 percent), outlet control will occur for both a submerged or unsubmerged inlet and a submerged or unsubmerged outlet. If the culvert is installed on a steep slope (3 percent), outlet control will occur if the tailwater is sufficiently high to submerge the critical depth at the upstream end of the culvert or to cause full barrel flow throughout. If a culvert is installed on a steep slope and only a portion of the barrel is submerged, a hydraulic jump will occur in the barrel but the headwater will still be inlet controlled. The controlling relationship for outlet control is an energy balance between sections 1 and 2, as illustrated in figure16, but now the two sections are separated by all the losses and elevation changes that occur between the tailwater channel and the approach channel. The outlet control head-discharge relationship, in which head relates to the total energy, is shown in the equation in figure 17.

$$ HW_o + \frac{V_u'^2}{2g} = TW + \frac{V_d'^2}{2g} + H_L $$

Figure 17. Equation. Head-discharge relationship, outlet control.

Where:

HW_o	is headwater depth above the outlet invert.
TW	is tailwater depth above the outlet invert.
V_u	is approach velocity.
V_d	is downstream velocity.
H_L	is total energy loss.

15

The total energy losses include the entrance, friction, and exit losses, as shown in figure 18.

$$H_L = H_{Le} + H_{Lf} + H_{Lo}$$

Figure 18. Equation. Total energy losses.

Where:

H_{Le}	is entrance loss.
H_{Lf}	is friction loss.
H_{Lo}	is exit loss.

The entrance and exit losses are commonly expressed as a fraction of the barrel velocity head, as shown in figures 19 and 20.

$$H_{Le} = K_e \left(\frac{V^2}{2g} \right)$$

Figure 19. Equation. Entrance loss.

$$H_{Lo} = K_o \left[\left(\frac{V^2}{2g} \right) - \left(\frac{V_d^2}{2g} \right) \right]$$

Figure 20. Equation. Exit loss.

Where:

K_e	is an entrance loss coefficient.
K_o	is an exit loss coefficient, usually assumed to be 1.0 for design purposes.
$V^2/2g$	is velocity head inside the culvert barrel.
$V_d^2/2g$	is velocity head in the downstream channel near the outlet, often neglected in design.

Since this study was focused on the effects of inlet geometry, the primary emphasis was on the entrance loss coefficient for outlet control, but exit losses were measured to determine if multiple barrels or skewed inlets affected the exit loss coefficient. It would be reasonable to express the entrance loss in terms of a difference in velocity heads, especially for wide culverts with low contraction ratios, but in this study the data were not analyzed that manner.

16

CHAPTER 4. DATA ACQUISITION AND DATA ANALYSIS PROCEDURES

This chapter describes the data acquisition system used for the culvert study and explains the data analysis procedures.

DATA ACQUISITION FOR CULVERT SETUP

The pressure sensors installed to measure the hydraulic grade line (HGL), the flow meters to gauge discharge, and the tailgate control for the culvert setup were linked to a National Instruments FieldPoint® system, a modular distributed input/output system. FieldPoint is designed for measurement, control, and data logging applications that require reliable, rugged systems involving diverse sensors and actuators located centrally or spread over large distances. FieldPoint also provides the flexibility to choose an open, standard networking technology such as Ethernet, serial, or wireless that best suits an application. At the FHWA Hydraulics Laboratory, wireless networking technology is used to integrate Fieldpoint into the lab data acquisition system. National Instruments' LabVIEW® graphical development software provides the tools to create measurement and control applications for FieldPoint.

The pressure sensors were Baumer Sensopress Type PCRD D015.14C.B110. The maximum pressure that can be applied to the sensor is 10 kPa (100 millibars or a 39-inch water column). For inlet control tests, only the tailbox pressure sensor readings were analyzed. For outlet control tests the pressure sensor readings in head, barrels, and tailbox were evaluated. To verify the pressure sensor data in the barrels, four standpipes per barrel were installed. Pressure sensors were mounted on the bottom of the standpipes. In addition to the pressure sensors, a scale attached to a side of the standpipes was used for manual readings of the water column in the standpipes. Scales were also mounted in the head and tailbox to validate the electronic readings.

DATA MANAGEMENT

Data management and data analysis were performed using the LabVIEW graphical programming technique for building applications such as testing and measurement, data acquisition, instrument control, data logging, measurement analysis, and report generation. LabVIEW programs are called virtual instruments (VIs) because their appearance and operation imitate physical instruments such as oscilloscopes and multimeters. Every VI uses functions that manipulate input from the user interface or other sources and displays that information or moves it to other files or other computers. Figure 21 shows the calculation procedure used to obtain inlet performance curves, inlet coefficients, and outlet coefficients.

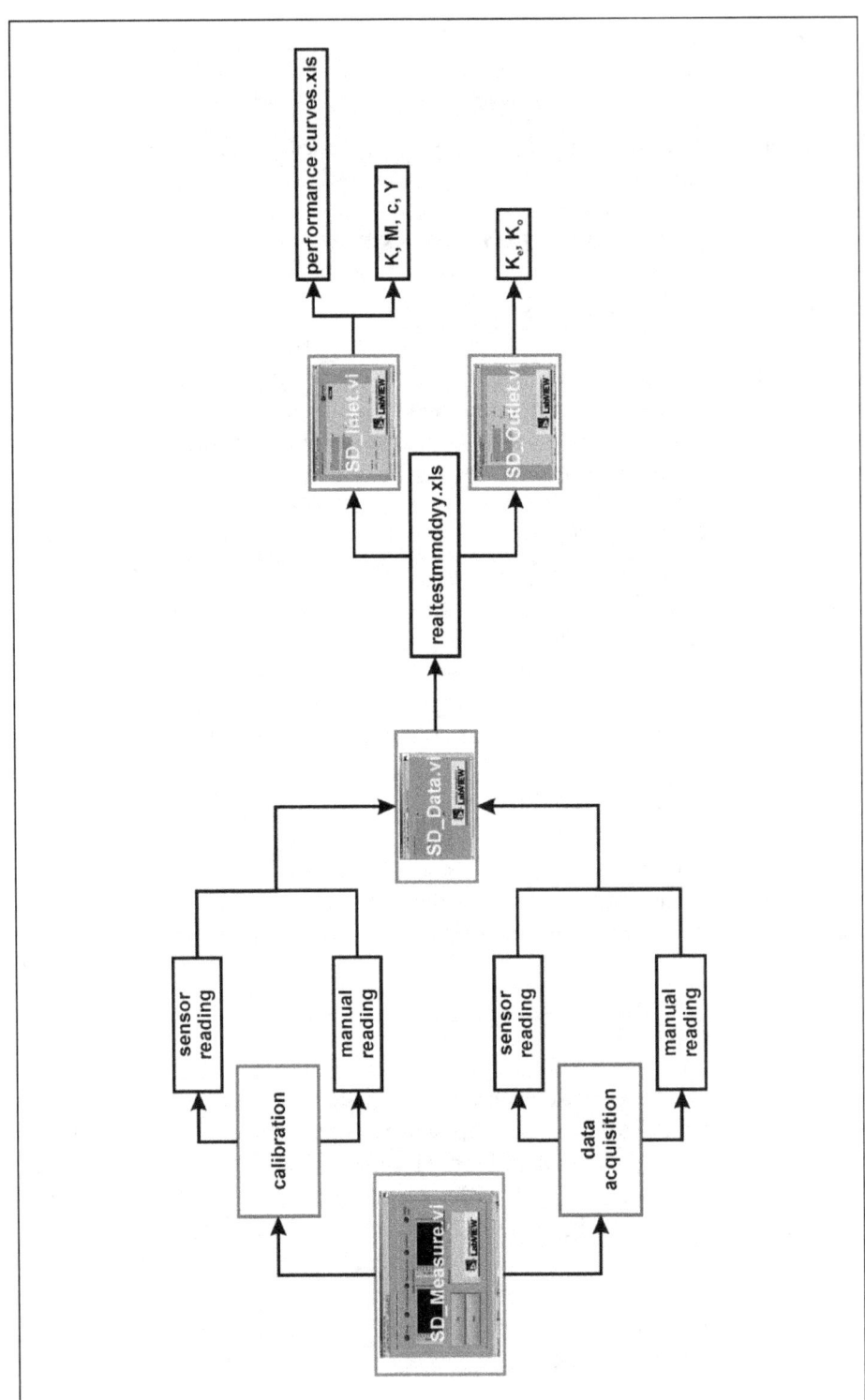

Figure 21. Diagram. Data management flow chart.

DATA ANALYSIS FOR INLET CONTROL TESTS

Using the equation in figure 14 in the preceding chapter as a regression equation, inlet control coefficients K and M for unsubmerged flow conditions were derived directly from performance curve data. For submerged flow, the terms c and Y were derived using the equation in figure 15 as a regression equation.

When a full sequence of trial data was collected (ideally five unsubmerged and five submerged trials), the regression analysis was done using the nonlinear Levenberg-Marquardt fit algorithm in LabVIEW to determine the set of coefficients that minimized the chi-squared quantity (figure 22).

$$\chi^2 = \sum_{i=0}^{N-1} \left(\frac{y_i - f(x_i; a_1, \ldots a_M)}{\sigma_i} \right)^2$$

Figure 22. Equation. Regression analysis, chi-squared.

In this equation, x_i and y_i are the input data points, $f(x_i; a_1, \ldots a_M)$ is the nonlinear function, and $a_1, \ldots a_M$ are coefficients. If the measurement errors are independent and normally distributed with a constant standard deviation $\sigma_i = \sigma$, the equation gives the least square estimation.

In addition to the proposed inlet coefficients (K, M, c, Y), fifth-order polynomials for inclusion in future updates to HDS-5 were derived to fit the unsubmerged and submerged data points (figure 23).

$$\frac{HW_i}{D} = a + b \left[\frac{Q}{AD^{0.5}} \right] + c \left[\frac{Q}{AD^{0.5}} \right]^2 + d \left[\frac{Q}{AD^{0.5}} \right]^3 + e \left[\frac{Q}{AD^{0.5}} \right]^4 + f \left[\frac{Q}{AD^{0.5}} \right]^5$$

Figure 23. Equation. Fifth-order polynomial for HW/D.

The coefficients a through f are the polynomial coefficients. The other terms have been previously defined.

The best-fit coefficients were calculated using LabVIEW's general polynomial fit virtual instrument, which is based on the least squares procedure to estimate the best fit.

DATA ANALYSIS FOR OUTLET CONTROL TESTS

Outlet control entrance loss is just one component that is added to friction and outlet losses to relate headwater elevations to tailwater (TW) elevations. Data from outlet loss experiments require careful scrutiny to avoid reporting unreasonable results. The entrance loss coefficient, K_e, for outlet control was computed from the relationship in figure 19 in the preceding chapter. Figure 19 is rearranged in figure 24.

$$H_{Le} = K_e \left(\frac{V^2}{2g} \right) \quad or \quad K_e = H_{Le} \div \left(\frac{V^2}{2g} \right)$$

Figure 24. Equation. Entrance loss coefficient.

H_{Le} is the entrance loss component that is usually computed for the design situation but is measured in the lab to compute K_e.

The technique used to measure H_{Le} illustrated in figure 25 involved extrapolating energy grade lines (EGLs) in the headbox and in the culvert barrel to a common plane and taking the difference in the EGLs at that common plane. Standard practice is to report an average design value of K_e that is a function of the inlet type only.

For higher discharge intensities, when the culvert barrel was at or near full, the computed K_e values were reasonably constant to warrant reporting an average value for design purposes. For the lower discharge intensities, when the culvert barrel was partly full throughout, the K_e values scattered considerably. This scatter was the result of the velocity head being very small and approaching zero for the very low discharges. In addition, H_{Le} was sensitive to the extrapolation process and to the flow distribution in the multiple barrels that pushed the resolution limits of the pressure transducers (figure 26). The difficulties of the extrapolation process for low flows justified splitting K_e into an unsubmerged and a submerged coefficient.

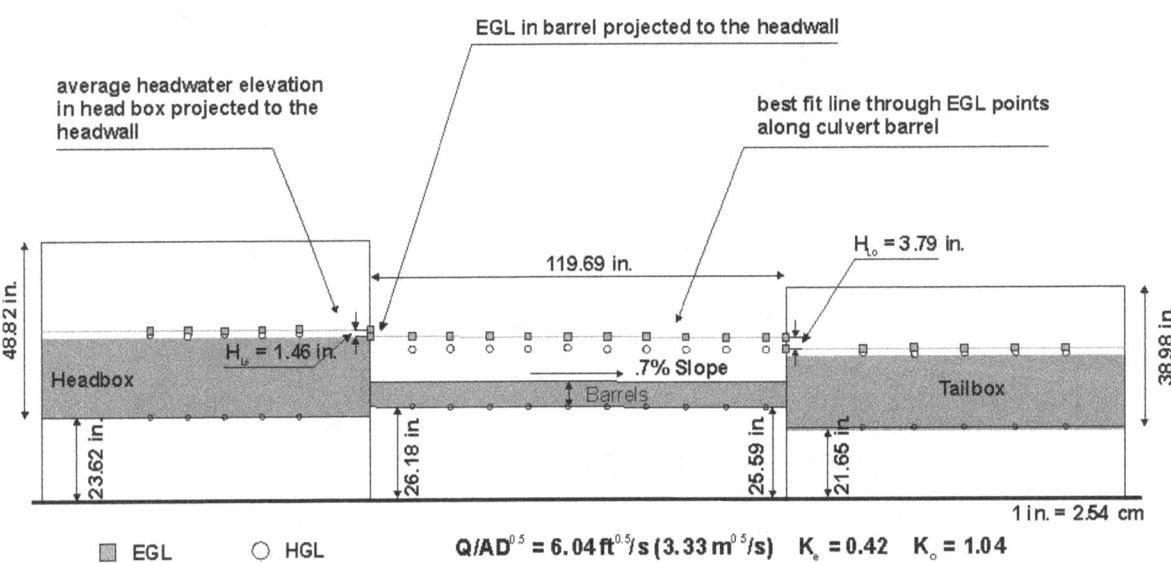

Figure 25. Diagram. Technique to determine H_{Le}.

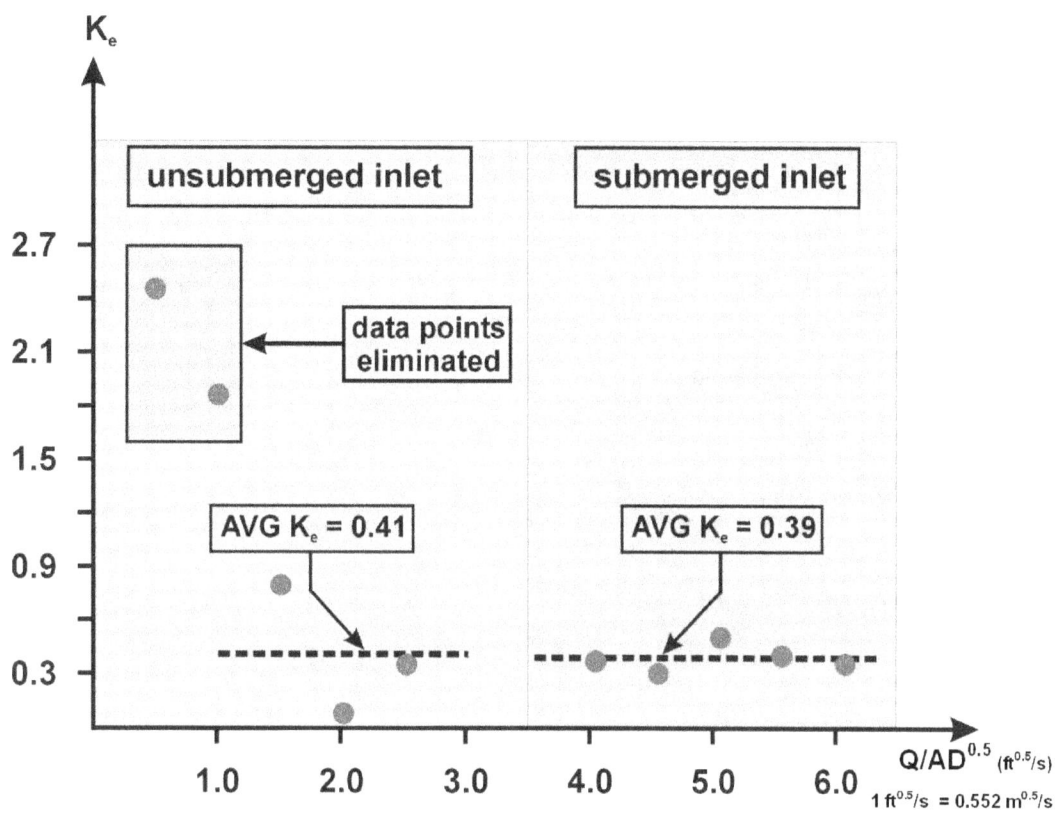

Figure 26. Graph. Typical behavior of K_e versus discharge intensity.

CHAPTER 5. EXPERIMENTAL PROCEDURES

MINIFLUME EXPERIMENTS

A miniflume was used to conduct experiments to optimize the bevel edges of the wing walls and top edges using two-dimensional PIV. The miniflume was 2800 millimeters (mm) (109.2 inches) long and 450 mm (17.6 inches) wide (figure 27). The culvert model and bevel models (figure 28) were constructed with Plexiglas to avoid reflections. The scale of the models was 1:30. The upstream flow conditioning was achieved using filter mats and honeycomb flow straighteners. The sidewalls of the flume were made of glass, allowing excellent flow visibility. The flow discharge could be varied between 0 and 5 liters per second (L/s) (0 and 1.32 gallons per second (gal/s)). An ultrasonic distance meter measured the flow depth. An electromagnetic velocity probe measured the approach velocity.

A two-dimensional PIV system typically consists of several subsystems. In most applications, tracer particles are added to the flow. These particles are illuminated in a plane of the flow at least twice within a short time interval. The light scattered by the particles is recorded either on a single frame or on a sequence of frames. The displacement of the particle images between the light pulses is determined using cross correlation techniques. In this study, the camera, a charge coupled device (ccd) camera, used for PIV recordings, was a Roper Scientific MEGAPLUS® Model ES1.0 digital camera. The camera was a stand-alone device connected to a frame grabber card (NI PXI 1422) by a PXI computer. To generate the light sheet, a double-pulsed Solo PIV 120 laser was used. The Solo PIV 120 laser is a compact, dual laser head system designed to provide a highly stable green light source for PIV applications.

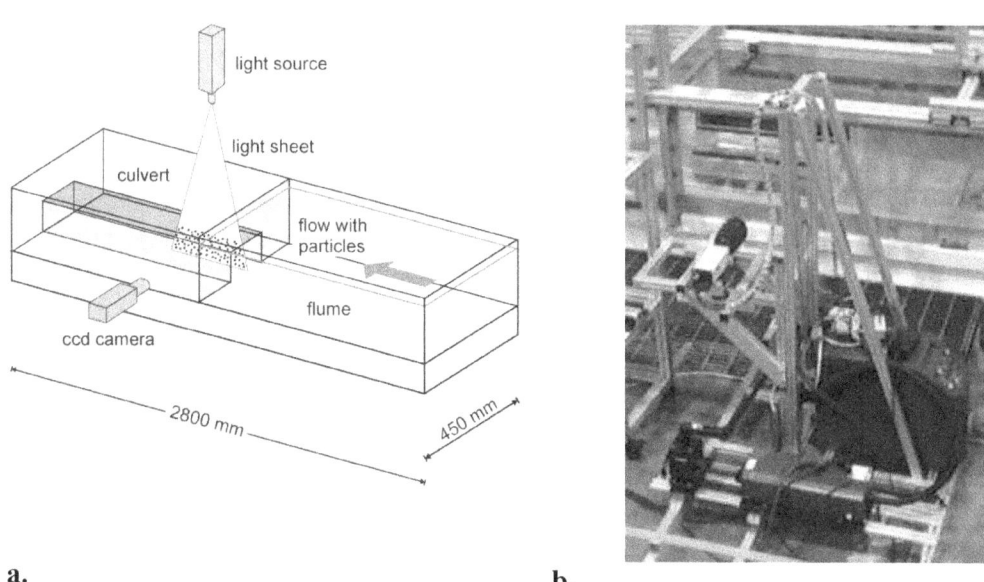

a.   b.

Figure 27. Diagram and Photo. Miniflume and PIV setup.

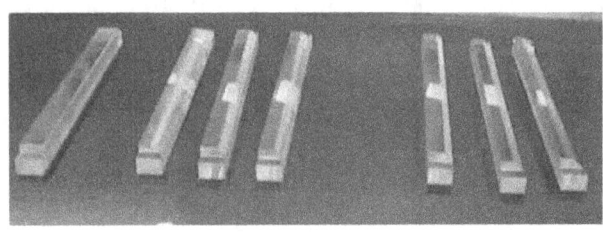 

a.   b.

Figure 28. Photos. Bevel models and PIV camera at culvert entrance.

PIV—Post Processing

The evaluation of the two-dimensional PIV data results in instantaneous velocity vectors of the flow field. To study the contracted area in the culvert, it was useful to integrate the velocity fields resulting in stream function and potential function. The integration is based on the assumption that the integrand, which is the flow field, is two-dimensional as well as incompressible. In this case, potential theory relates the velocity field, $U = (U (X,Y), V (X,Y))$, to the stream function Ψ, which can be integrated over the domain (X, Y plane) as shown in figure 29. Fifty double-image frames were recorded to correlate 50 velocity flow field samples. The integration procedure for the stream functions was applied by taking the average of the velocity flow field sequence.

Seven different bevel edge conditions were tested for the miniculvert setup (figure 30). The criterion to determine the best bevel performance was the contracted distance outside the viscous boundary layer (effective flow depth at vena contracta).

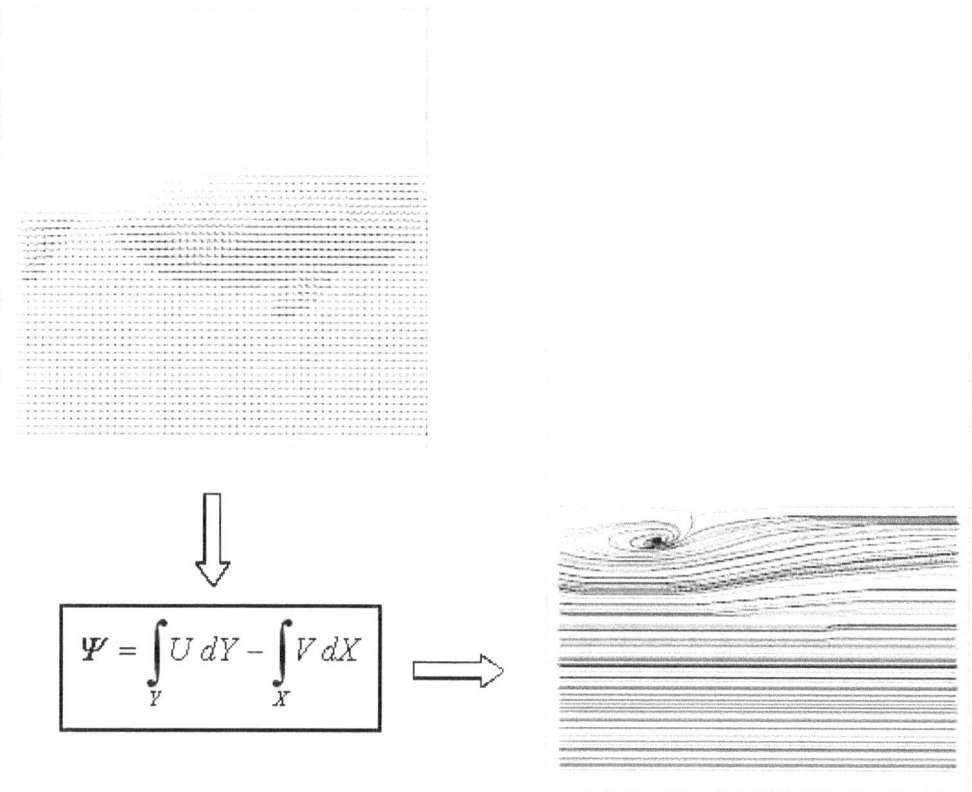

Figure 29. Diagram. Integration of velocity flow field in stream functions to study culvert flow contraction.

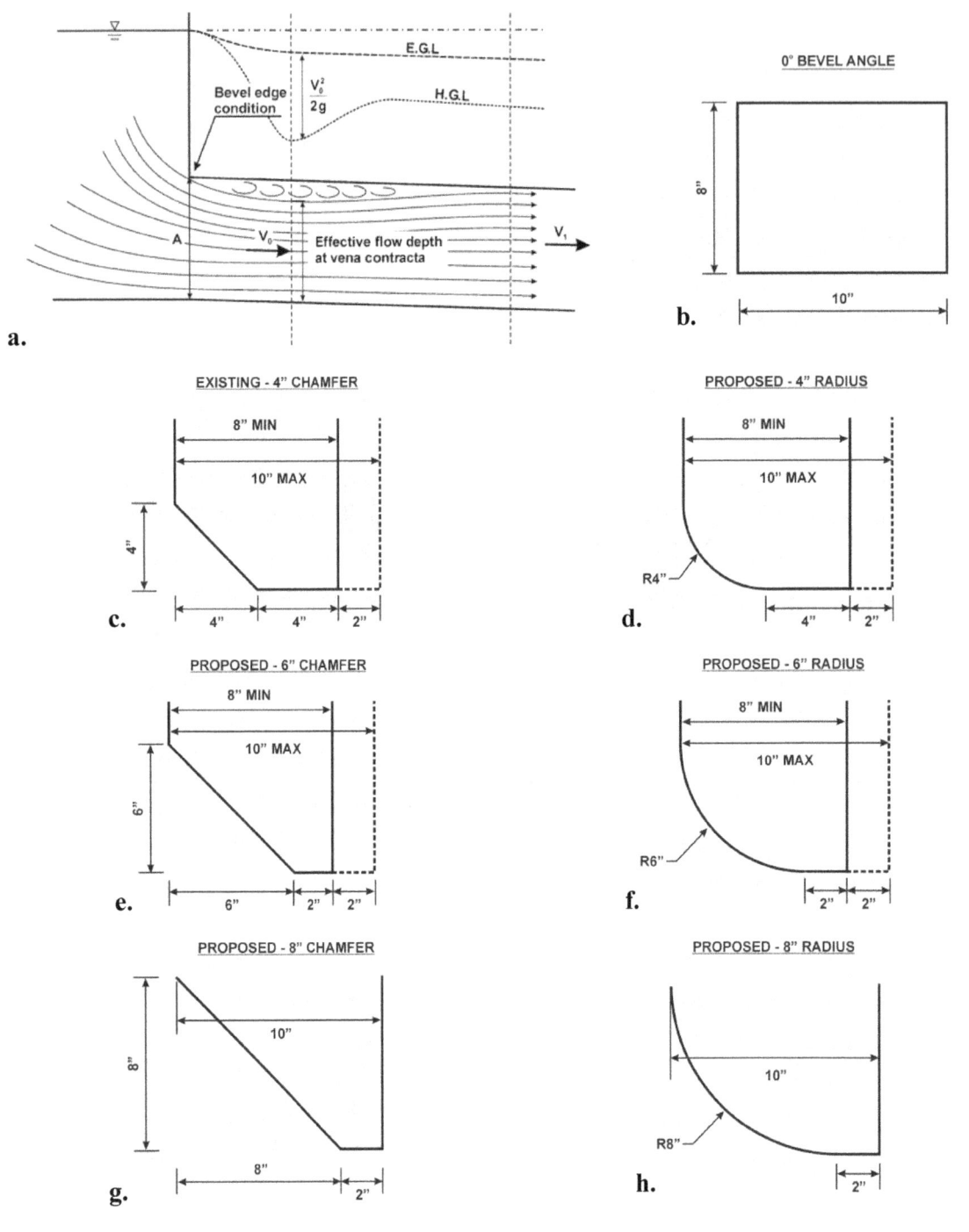

1 in. = 2.54 cm

Figure 30. Diagrams. Tested bevel edge conditions and effective flow depth criterion.

26

CULVERT SETUP EXPERIMENTS

The culvert setup consists of two water tanks (a headbox 2.44 m (8 ft) long by 2.44 m (8 ft) wide by 1.22 m (4 ft) high and a tailbox 2.44 m (8 ft) long by 1.83 m (6 ft) wide by 0.92 m (3 ft) high) that were connected to the tested culvert barrel (figures 31 to 34). Five electronic pressure sensors were integrated into the bottom floor of the head and tailbox. An additional 40 pressure sensors measured the hydraulic grade line inside the culvert model barrels. The side rails of the tailbox supported a two-dimensional robot to measure the velocity distribution inside the tailbox (figure 35). The robot was automated to measure an area perpendicular to the main flow direction. A laser distance sensor recorded the position of the robot in the flow direction. An automated tailgate at the downstream end of the tailbox allowed for adjustment of the tailwater depth. The culvert setup was fully automated (pump/flowmeter and tailgate/pressure sensor control logic) and was network controlled.

The experimental setup was based on flow being in alignment with the barrels. The barrels were tested for two different slope settings (3 percent and 0.7 percent). The purpose of the steep slope (3 percent) was to guarantee inlet control conditions. The flat (0.7 percent) slope was designed to simulate outlet control conditions and was controlled by the adjustable tailgate. The pressure sensors in the culvert barrel were evaluated for subcritical flow for each outlet control test run. If this was not the case, the tailgate was lowered until outlet control conditions were achieved. The velocity distribution at the exit of the culvert was also measured for each outlet control test run using an electromagnetic velocity probe mounted on a two-dimensional robot in the tailbox. Conclusions based on the recorded velocity distribution were made about the flow distribution in the barrels.

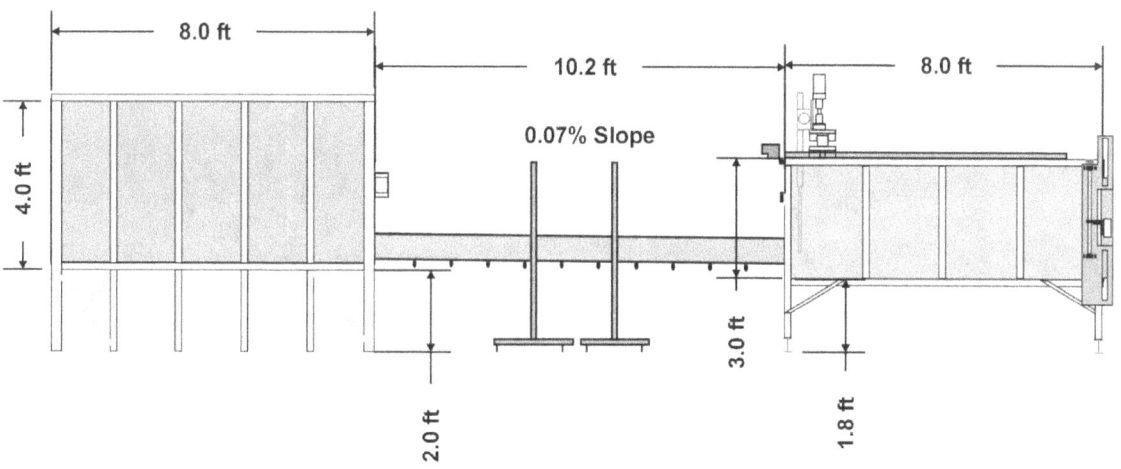

1 ft = 0.305 m

Figure 31. Diagram. Culvert setup—side view.

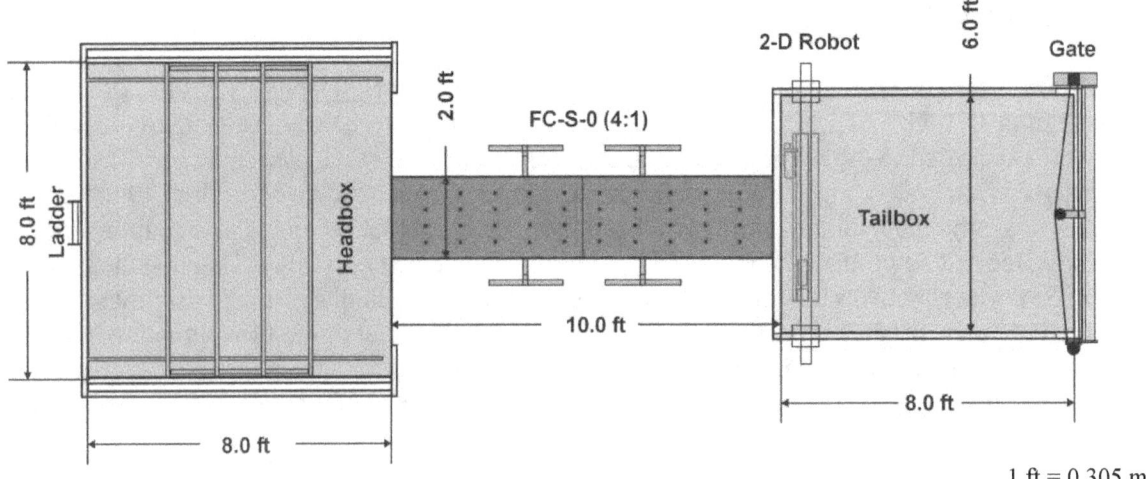

Figure 32. Diagram. Culvert setup—top view.

1 ft = 0.305 m

Figure 33. Photo. Culvert setup—overview.

Figure 34. Photo. Culvert model barrels.

Figure 35. Photo. Two-dimensional robot to measure velocity distribution in tailbox.

Inlet and Culvert Barrel Models

All barrels used for the tests were fabricated out of Plexiglas consisting of two 155.58-cm (61.25-inch) sections. Tapped holes for the pressure sensors were placed every 28.78 cm (11.33 inches) on the bottom side of the culvert model barrels. Special holes on the top and bottom of the barrels provided a mounting for the corner fillets. Two sets of model fillets (1.27 and 2.54 cm (0.5 and 1 inch)) were fabricated using polyvinyl chloride (PVC) material.

The inlet models were designed to consist of modular parts (wingwalls, center walls, top plates) that were easily changed for various configurations. Tongue and groove connectors were used to assemble the models (figure 36).

Overviews of models tested in the culvert setup are listed in appendix A.

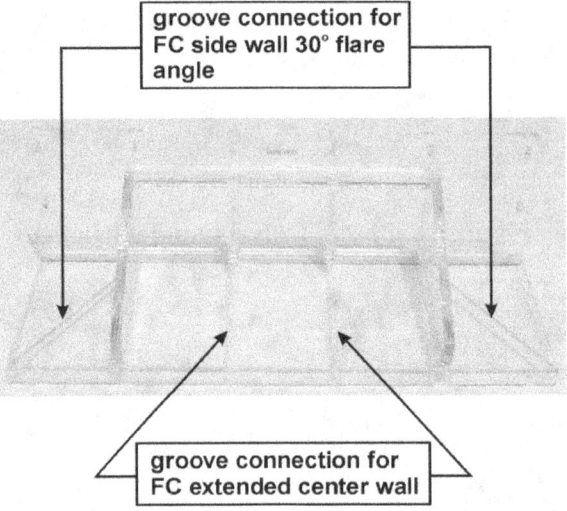

Figure 36. Photo. Groove connectors to assemble models.

CHAPTER 6. EXPERIMENTAL RESULTS

EFFECTS OF BEVELS AND CORNER FILLETS

Miniflume Test Results for Bevel Effects

The small demonstration miniflume was used with PIV to determine the best shape for the top edge from several alternate shapes that were suggested. The goal of the PIV miniflume experiments was to accurately measure the flow field at the vena contracta. Post processing of PIV results provides streamlines that can be visually interpreted to show the shape that produces the maximum effective flow depth at the vena contracta. This shape is likely to have the least headloss when incorporated into the inlet geometry. The effective flow depth at the vena contracta for various bevel edge conditions are shown in figures 37 and 38.

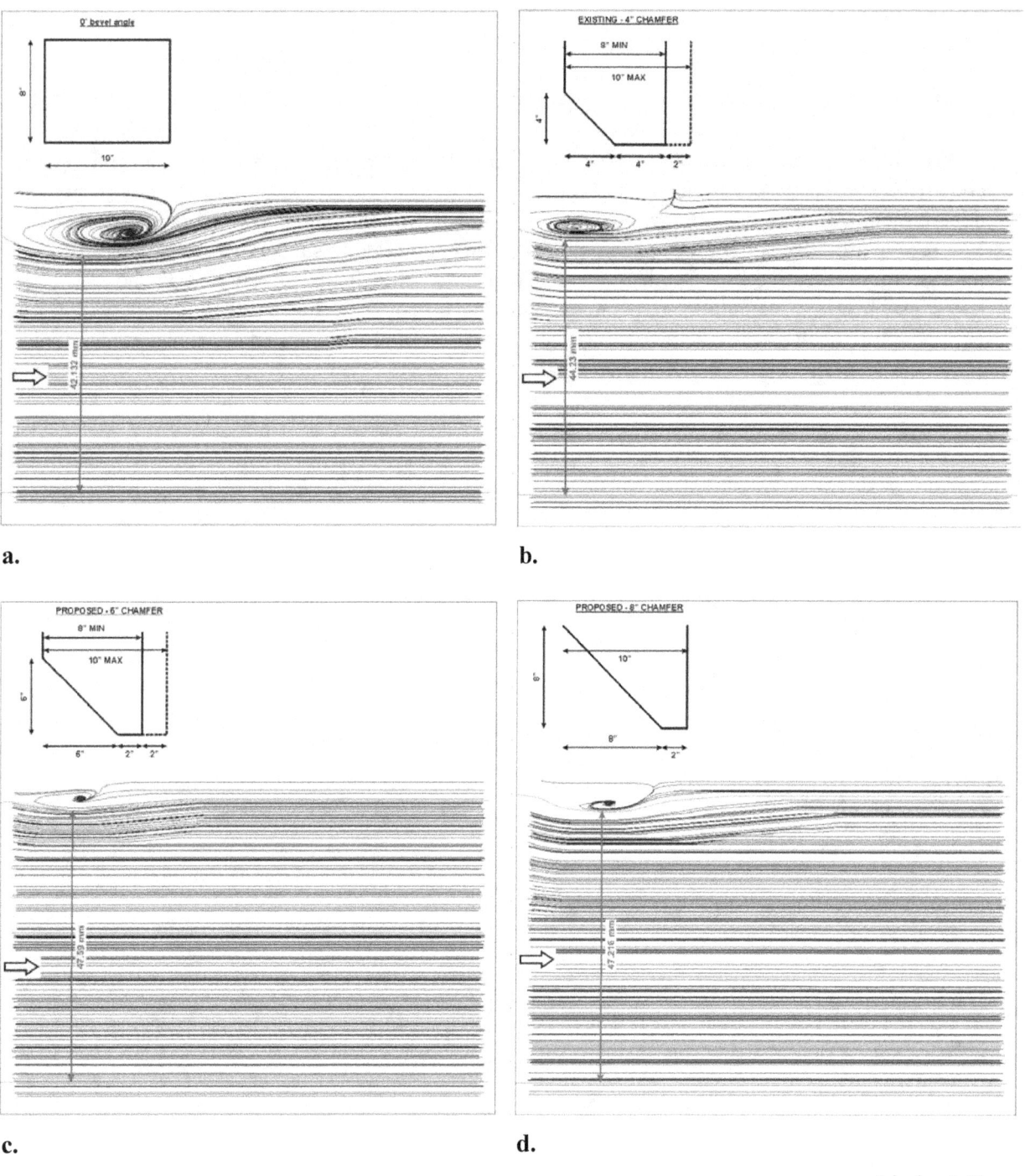

a.

b.

c.

d.

1 inch = 2.54 cm

Figure 37. Diagrams. Effective flow depth at vena contracta for nonrounded bevel edges.

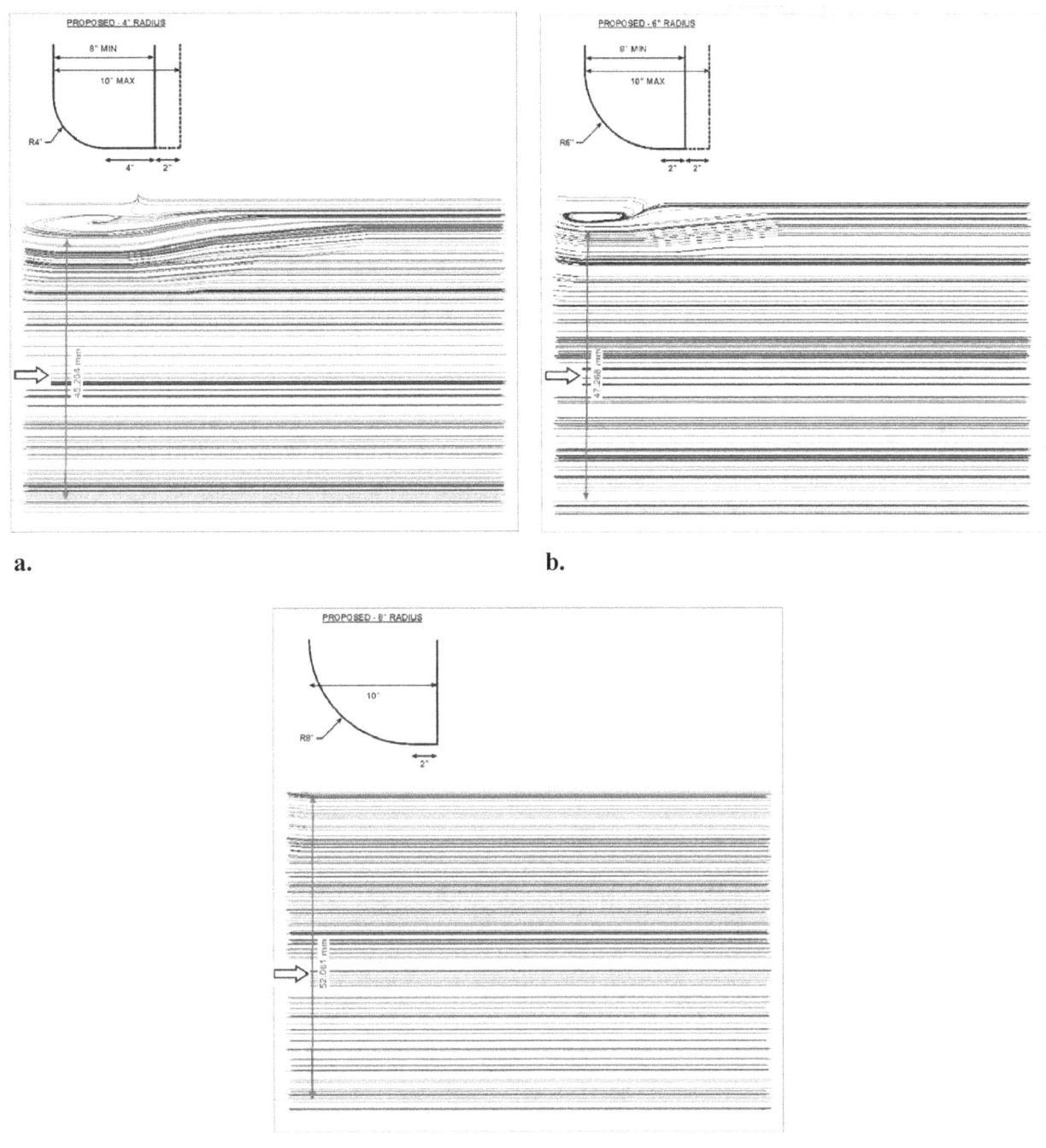

a.

b.

c.

1 inch = 2.54 cm

Figure 38. Diagrams. Effective flow depth at vena contracta for rounded bevel edges.

For each experiment, headwater and tailwater depths were measured. The difference of these depths versus the vena contracta measurement is plotted on figure 39. The results show that the 20.32-cm- (8-inch-) rounded bevel edge produces the optimal geometric configuration.

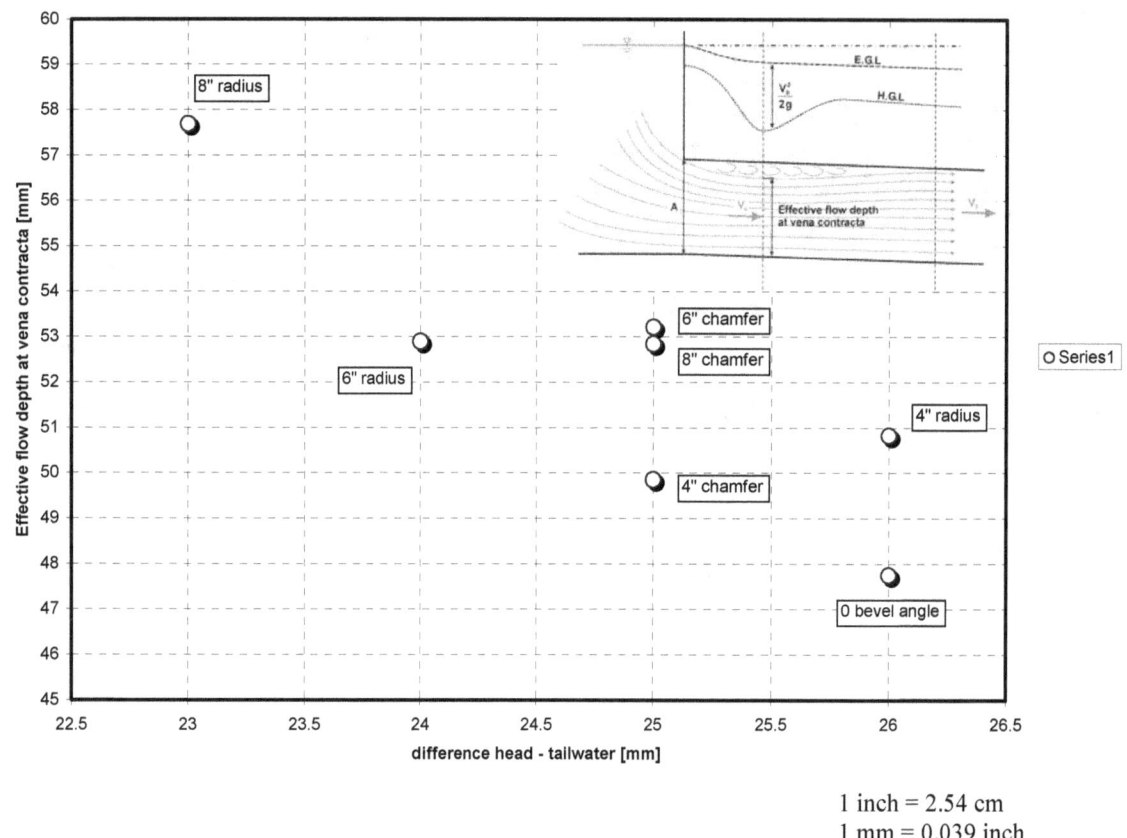

1 inch = 2.54 cm
1 mm = 0.039 inch

Figure 39. Graph. Effective flow depth versus headwater/tailwater difference.

The PIV miniflume tests demonstrated which shape is best, but the question then becomes how much gain does this shape produce in the hydraulic performance? To answer that question, tests were conducted on the effects of bevels and of corner fillets. Sketches and descriptions of the models tested are set forth in figure 40. The results are discussed in the remainder of this section, and selected information is presented in table 1 at the end of the section.

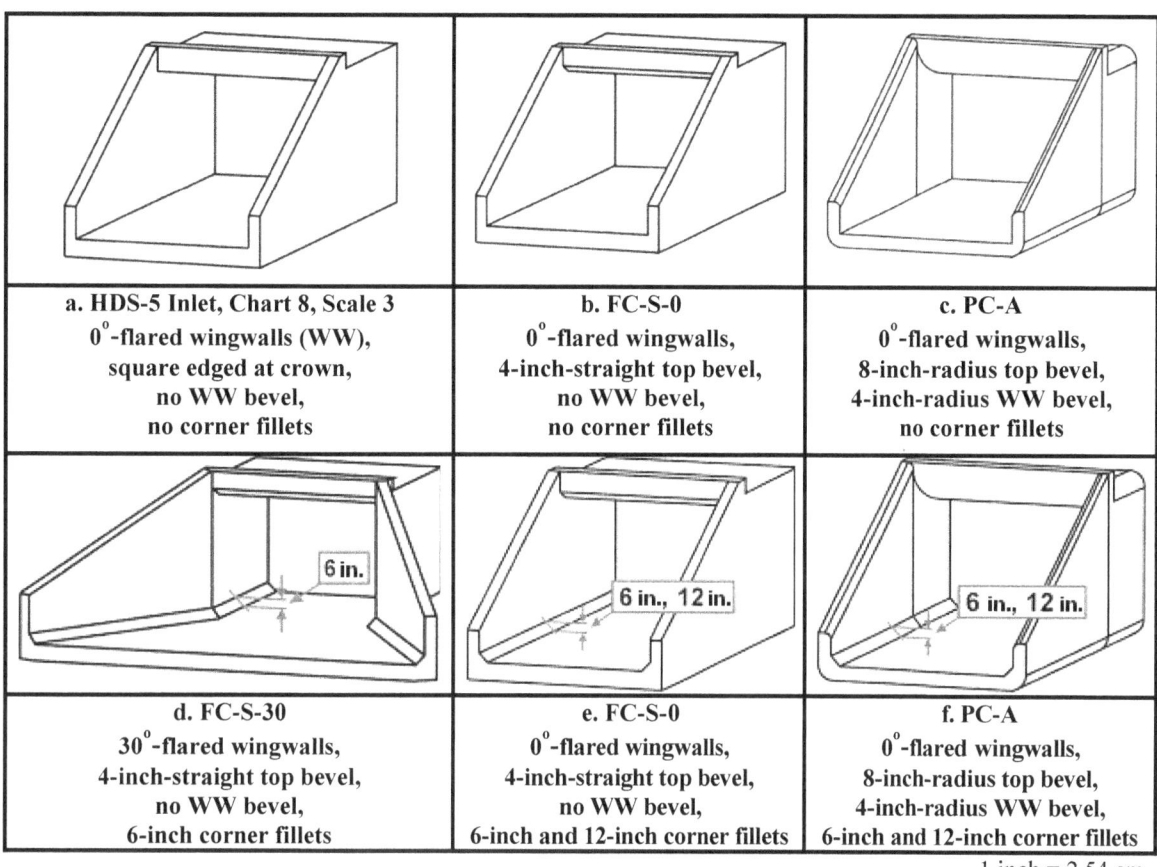

a. HDS-5 Inlet, Chart 8, Scale 3 0°-flared wingwalls (WW), square edged at crown, no WW bevel, no corner fillets	b. FC-S-0 0°-flared wingwalls, 4-inch-straight top bevel, no WW bevel, no corner fillets	c. PC-A 0°-flared wingwalls, 8-inch-radius top bevel, 4-inch-radius WW bevel, no corner fillets
d. FC-S-30 30°-flared wingwalls, 4-inch-straight top bevel, no WW bevel, 6-inch corner fillets	e. FC-S-0 0°-flared wingwalls, 4-inch-straight top bevel, no WW bevel, 6-inch and 12-inch corner fillets	f. PC-A 0°-flared wingwalls, 8-inch-radius top bevel, 4-inch-radius WW bevel, 6-inch and 12-inch corner fillets

1 inch = 2.54 cm

Figure 40. Sketches. Models tested for effects of bevels and corner fillets.

Head Loss Experiments for Bevel Effects

Precast (PC) models, fabricated with the optimum bevel on the top plate and 10.16-cm- (4-inch-) radius rounded bevels on the wingwall edges, and field cast (FC) models, fabricated with the SDDOT standard straight bevel on the top plate and no bevels on the wingwall edges, were tested in the culvert test facility. Additional tests were made with a model of the closest HDS-5 inlet that has 0-degree-flared wingwalls and no bevels on either the top plate or the wingwalls and with a model of an FC 30-degree-flared wingwall culvert. For unsubmerged flow when the top plate is not exposed to the flow, figures 41 and 42 show that the performance curve for the PC model with the optimum top plate bevel and rounded wingwall edges is almost identical to the performance curve for the comparable FC model with 0-degree wingwall flare but with the straight top plate bevel and square edge wingwall edges. That comparison suggests that the 10.16-cm- (4-inch-) radius bevel on the wingwall edges does not contribute much gain. The gain shows up in the submerged condition with approximately 10 percent headwater reduction at the highest discharge intensities. Figure 43 shows that the optimum bevel model is almost as efficient as the 30-degree-flared wingwall at the higher discharge intensities.

35

The performance curves shown in figures 41 and 42 compare inlets and barrels with the same corner fillets to isolate the effects of the bevels. In South Dakota, the PC culverts are fabricated with 30.48-cm (12-inch) corner fillets and FC culverts are constructed with 15.24-cm (6-inch) corner fillets, but those sizes were varied in this study to make clear comparisons.

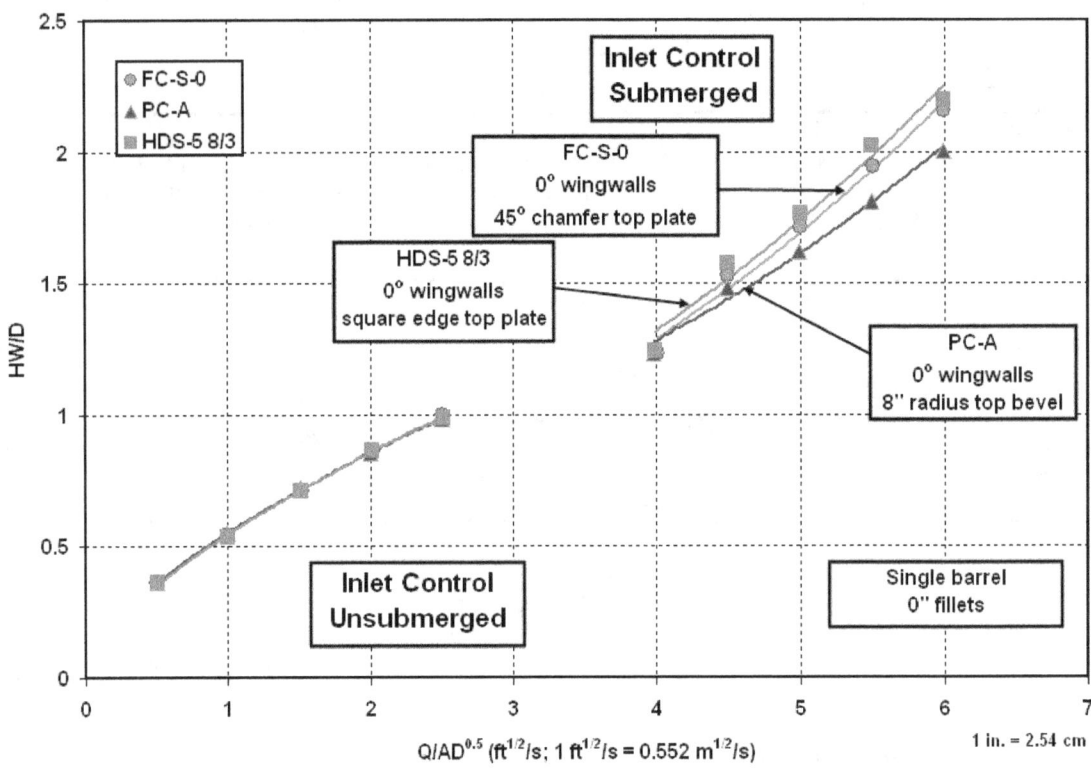

Figure 41. Graph. Inlet control performance curves, FC-S-0 versus PC-A, zero corner fillets.

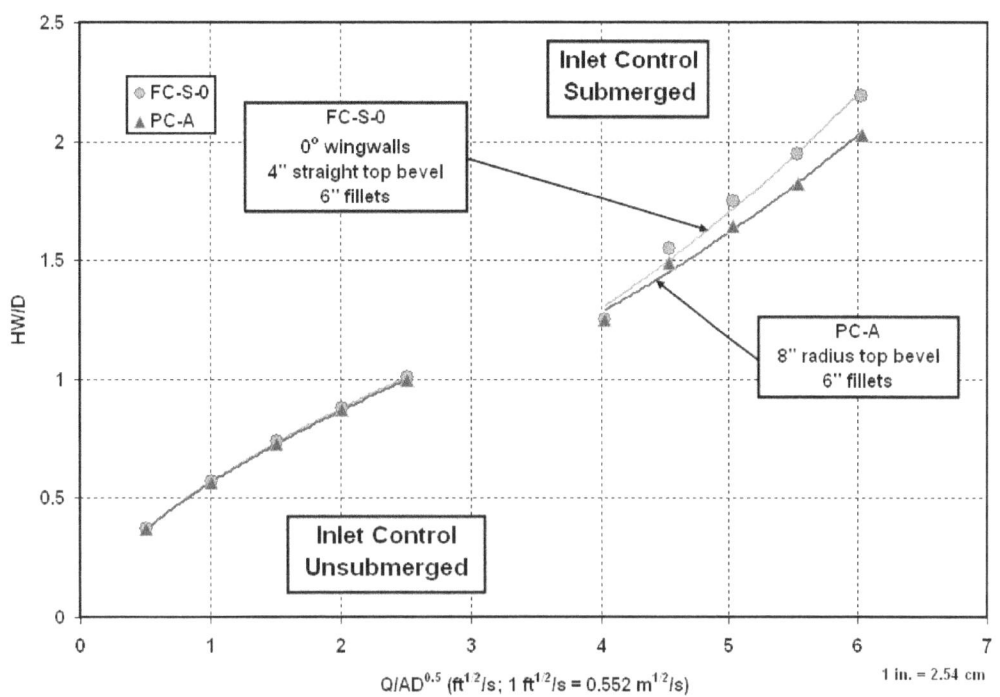

Figure 42. Graph. Inlet control performance curves, FC-S-0 versus PC-A, 15.24-cm (6-inch) fillets.

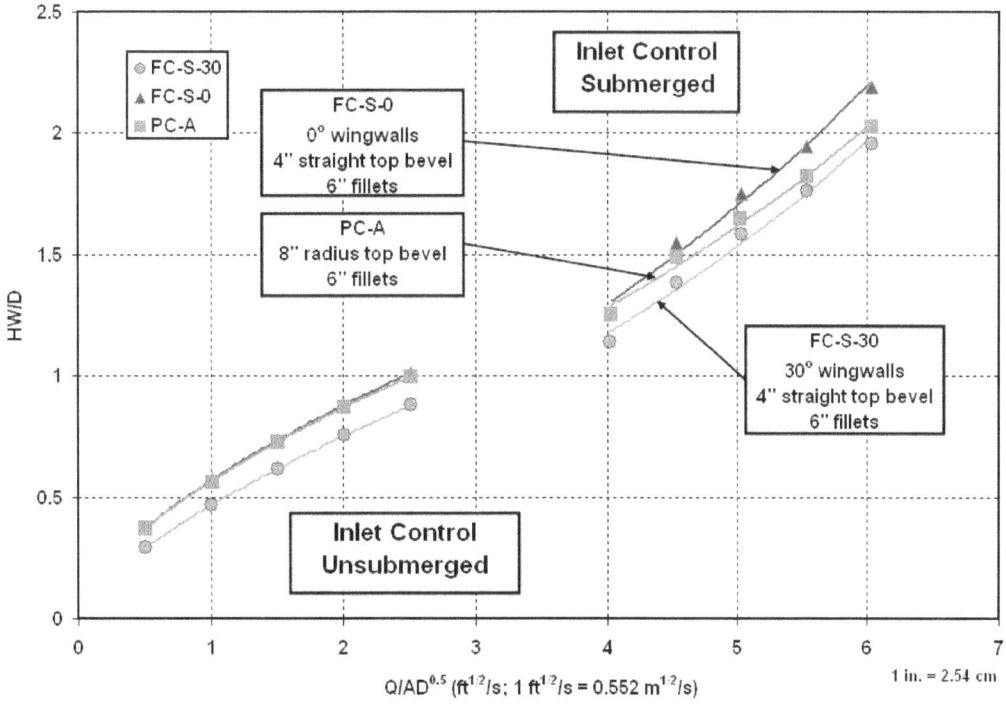

Figure 43. Graph. Inlet control, precast with 15.24-cm (6-inch fillets) and field cast with 15.24-cm (6-inch) fillets.

Effect of Wingwall Top Edge Bevel

Tests were conducted to isolate the effects of the bevels on the edges of the wingwalls. Earlier comparisons suggested that the wingwall bevels had very little effect, but those comparisons were based on the unsubmerged flow only.

Since the models were precisely fabricated with clip-on components, it was relatively easy to mix and match components to isolate the wingwall bevels. Two special models were tested. One was the FC model top plate with the PC model wingwalls and is labeled "FC-hybrid" in figure 44. The other was the PC top plate with the FC wingwalls and is labeled "PC-hybrid" in figure 45. In both cases, the performance curves for the hybrid models plotted almost identically over the curves for the models with the other wingwalls for both unsubmerged and submerged flow. The slight deviation in figure 45 is on the wrong side to suggest that there is any advantage to rounding the wingwall edges.

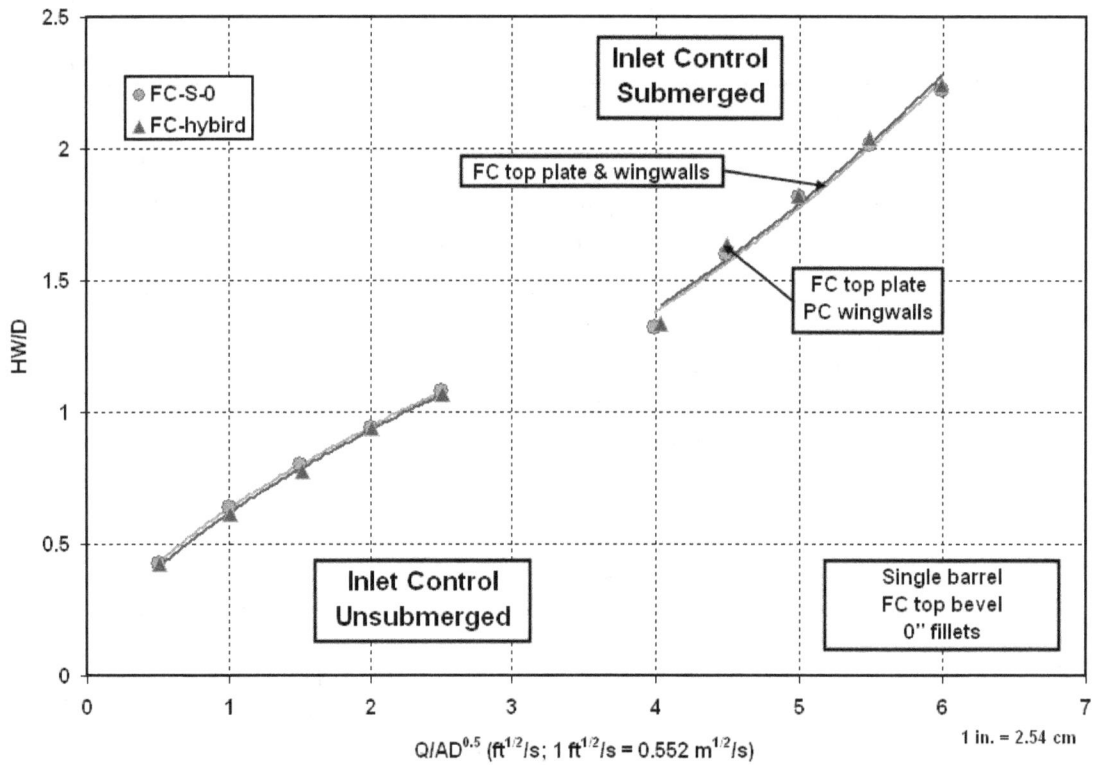

Figure 44. Graph. Inlet control, field cast hybrid inlet with 10.16-cm- (4-inch-) radius bevel on wingwalls.

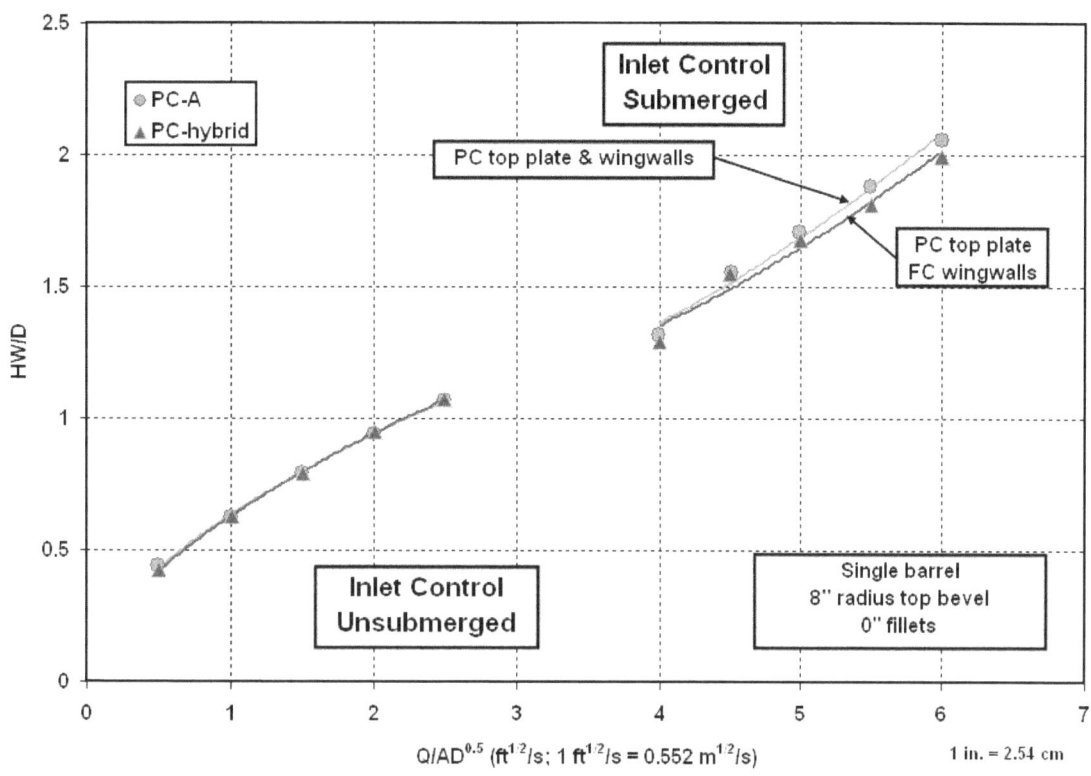

Figure 45. Graph. Inlet control, precast hybrid inlet with no bevel on wingwalls.

Effect of Corner Fillets

Corner fillets are fabrication expedients intended to minimize high-stress areas in the corners for rectangular culvert shapes. The PC industry tends to use slightly larger corner fillets than might be used for FC construction. Obviously, the corner fillets reduce flow area slightly, but the issue is what effect they might have on hydraulic performance beyond that reduction in flow area. Varying the corner fillets was one of the most aggravating challenges in the experimental design for this study. Ideally, the experimental coefficients would have been the same for the different fillets provided the correct net area was used in the computations. Then the same corner fillets could have been used for all of the tests and the culvert models would have been much simpler to fabricate.

Even though the net area was used in computations, the entrance loss coefficients for outlet control did vary with the size of the corner fillets. The performance curves for inlet control show no difference using 0-cm (0-inch), 15.24-cm (6-inch), or 30.48-cm (12-inch) corner fillets. Figures 46 and 47 for FC and PC models, respectively, indicate that, for inlet control, there was no headwater increase with fillet size for any given discharge intensity. The entrance loss coefficients for outlet control increased significantly for the 30.48-cm (12-inch) fillets when compared with the 15.24-cm (6-inch) fillets and no fillets.

39

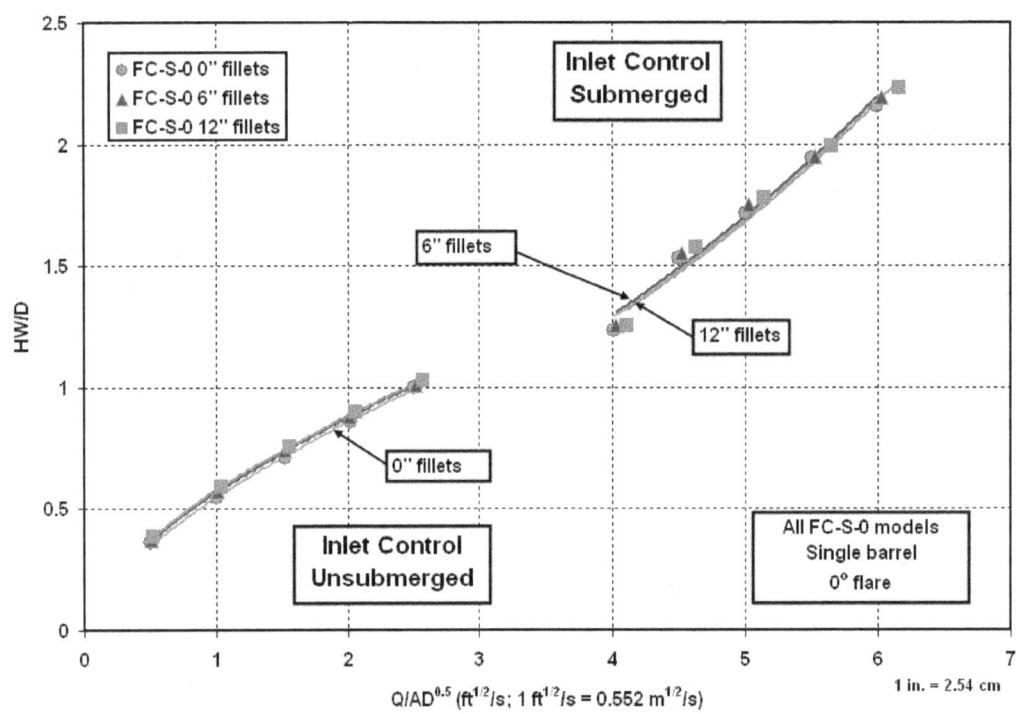

Figure 46. Graph. Inlet control effects of corner fillets for the field cast model.

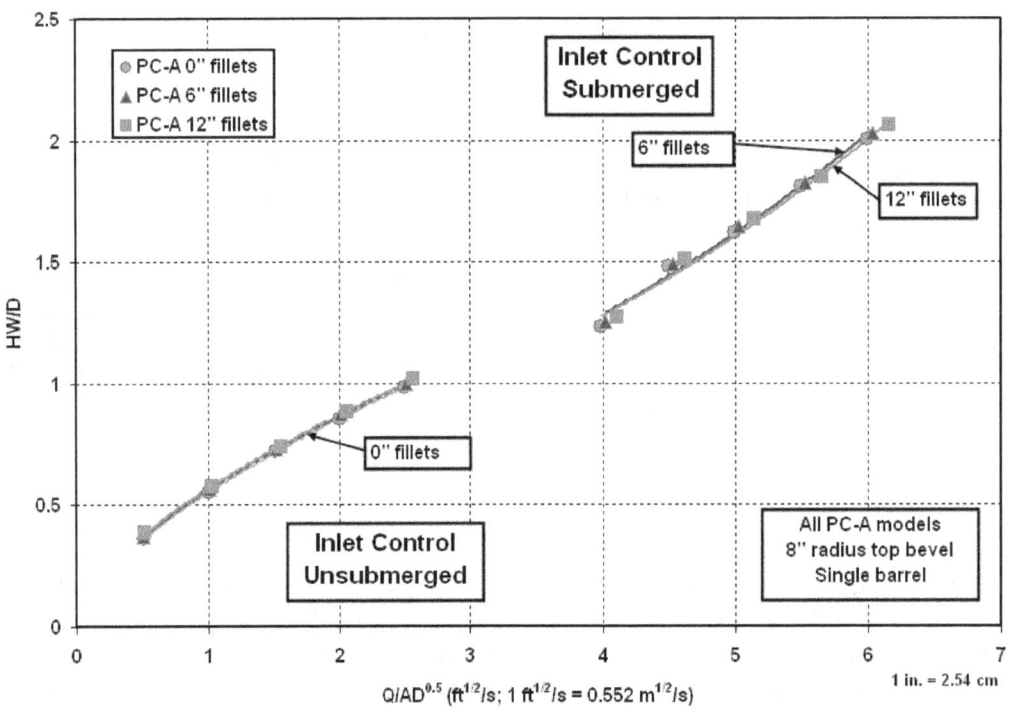

Figure 47. Graph. Inlet control effects of corner fillets for the precast model.

40

The likely combinations are probably 30.48-cm (12-inch) fillets with PC culverts and 15.24-cm (6-inch) fillets with FC culverts. Figure 48 compares the PC model with 30.48-cm (12-inch) corner fillets to the FC models with 15.24-cm (6-inch) fillets.

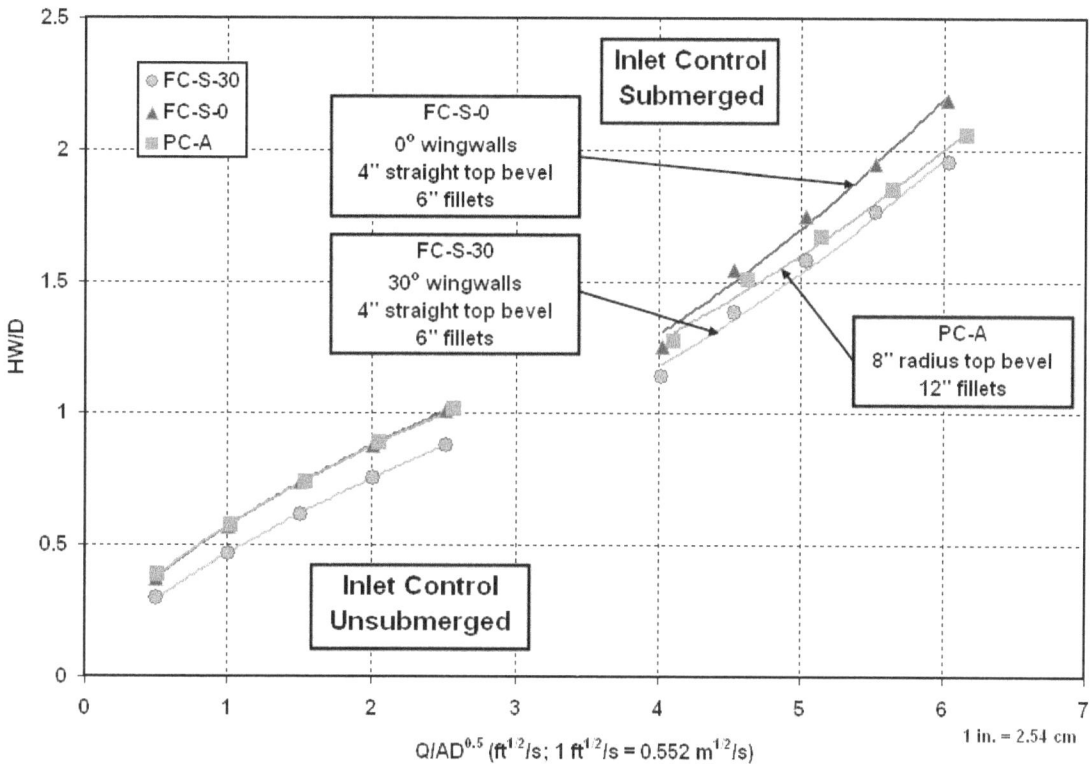

Figure 48. Graph. Inlet control, precast with 30.48-cm (12-inch) fillets and field cast with 15.24-cm (6-inch) fillets.

The entrance-loss coefficients, K_e, for outlet control, and the regression coefficients for the inlet control results plotted in figures 41 through 48 are summarized in table 1. The outlet control coefficient, K_e, for the PC model with the optimum bevels varied from 0.23 to 0.33; for the FC-S-0 model, it varied from 0.46 to 0.64; and for the FC-S-30 model with 30-degree-flared wingwalls, it was 0.26. Coefficients from different sources for the HDS-5 model are listed for comparison. The HDS-5 model presumably had square edges on the top plate as well as on the wingwalls.

Table 1. Effects of bevels and corner fillets—summary of inlet and outlet control coefficients.

Model	Span:Rise	Slope (percent)	Fillets (inches)	K_e	K Form 1 Eq.	M Form 1 Eq.	K Form 2 Eq.	M Form 2 Eq.	c	Y
FC-S-0	1:1	0.7	0	0.46						
		0.7	6	0.50						
		0.7	12	0.62						
		3	0	0.45			0.55	0.64	0.0453	0.54
		3	6	0.47			0.57	0.62	0.0448	0.56
		3	12	0.64			0.58	0.61	0.0447	0.54
PC-A	1:1	0.7	0	0.27						
		0.7	6	0.25						
		0.7	12	0.33						
		3	0	0.25			0.56	0.63	0.0371	0.67
		3	6	0.23			0.57	0.62	0.0371	0.67
		3	12	0.30			0.57	0.61	0.0361	0.68
PC-A	2:1	3	6	0.35			0.60	0.56	0.0329	0.79
PC-A	2:1	3	12	0.40			0.60	0.55	0.0316	0.81
FC-S-30	1:1	0.7	6	0.26						
HDS-5 chart 8 scale 3			0	0.70	0.061	0.75			0.0423	0.82
HDS-5 from Graziano (1996)	1:1	3	0	0.68	0.009	1.54	0.48	0.73	0.0435	0.62
HDS-5 from this study	1:1	3	0		0.0481	0.76	0.55	0.64	0.0469	0.55
		0.7	0	0.79						

1 inch = 2.54 cm.

Notes: For empty cells, data were not available or not applicable. Eq. is equation. Form 1 and form 2 equations are identified in chapter 3.

EFFECTS OF MULTIPLE BARRELS

Three questions need to be addressed with regard to multiple barrel culverts:
- Is it reasonable to assume that single barrel coefficients are applicable for multiple barrel culverts?
- Does the advantage of flared wingwalls diminish as the number of barrels or overall span-to-rise ratio increases? Is it less important to have flared wingwalls as the number of barrels increases?
- Is there a hydraulic advantage to extending the center walls of a multiple barrel culvert onto the approach apron?

Sketches of all the models tested in this series are illustrated in figure 49.

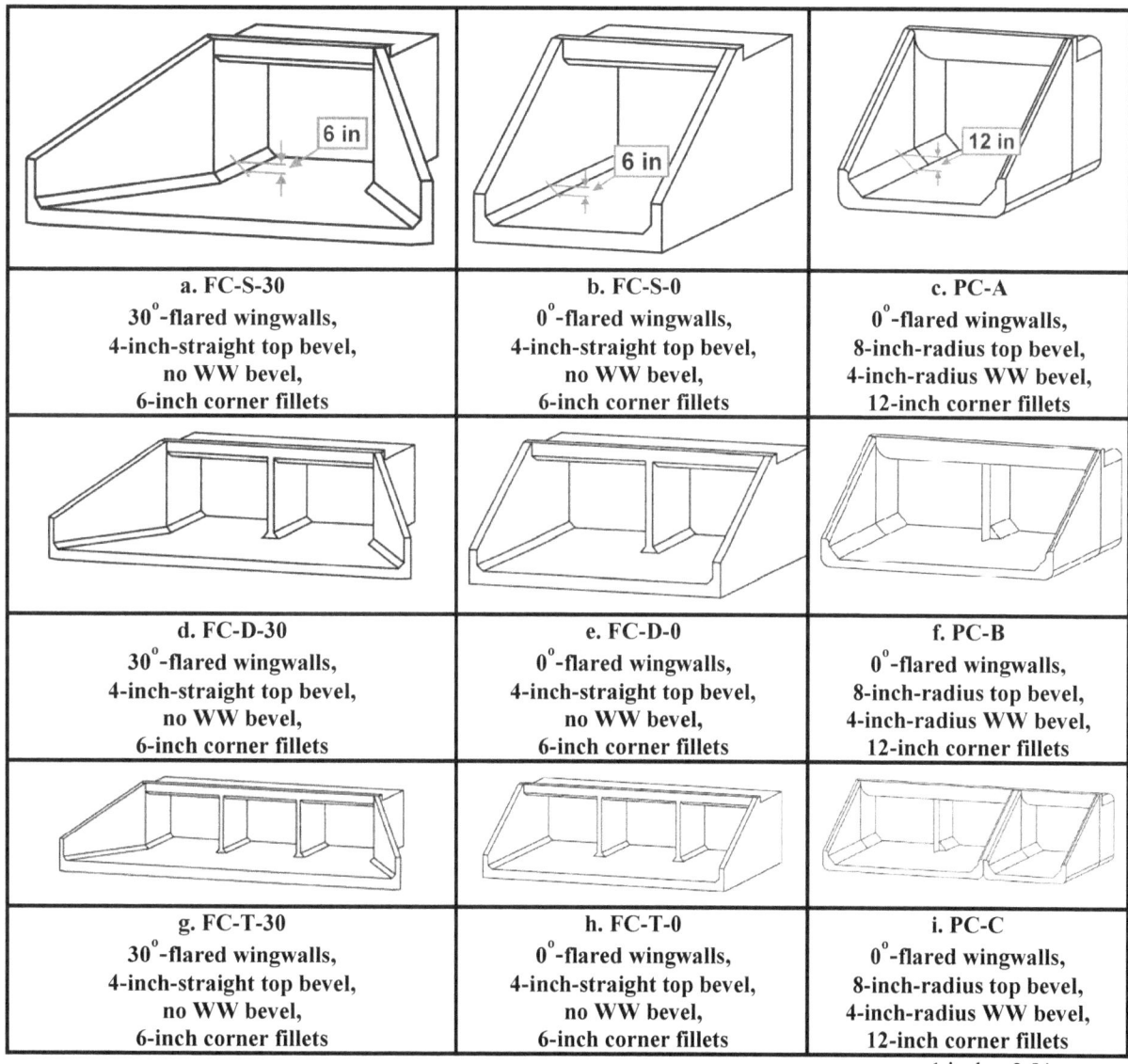

a. FC-S-30 30°-flared wingwalls, 4-inch-straight top bevel, no WW bevel, 6-inch corner fillets	b. FC-S-0 0°-flared wingwalls, 4-inch-straight top bevel, no WW bevel, 6-inch corner fillets	c. PC-A 0°-flared wingwalls, 8-inch-radius top bevel, 4-inch-radius WW bevel, 12-inch corner fillets
d. FC-D-30 30°-flared wingwalls, 4-inch-straight top bevel, no WW bevel, 6-inch corner fillets	e. FC-D-0 0°-flared wingwalls, 4-inch-straight top bevel, no WW bevel, 6-inch corner fillets	f. PC-B 0°-flared wingwalls, 8-inch-radius top bevel, 4-inch-radius WW bevel, 12-inch corner fillets
g. FC-T-30 30°-flared wingwalls, 4-inch-straight top bevel, no WW bevel, 6-inch corner fillets	h. FC-T-0 0°-flared wingwalls, 4-inch-straight top bevel, no WW bevel, 6-inch corner fillets	i. PC-C 0°-flared wingwalls, 8-inch-radius top bevel, 4-inch-radius WW bevel, 12-inch corner fillets

1 inch = 2.54 cm

Figure 49. Sketches. Models tested for the effects of multiple barrels.

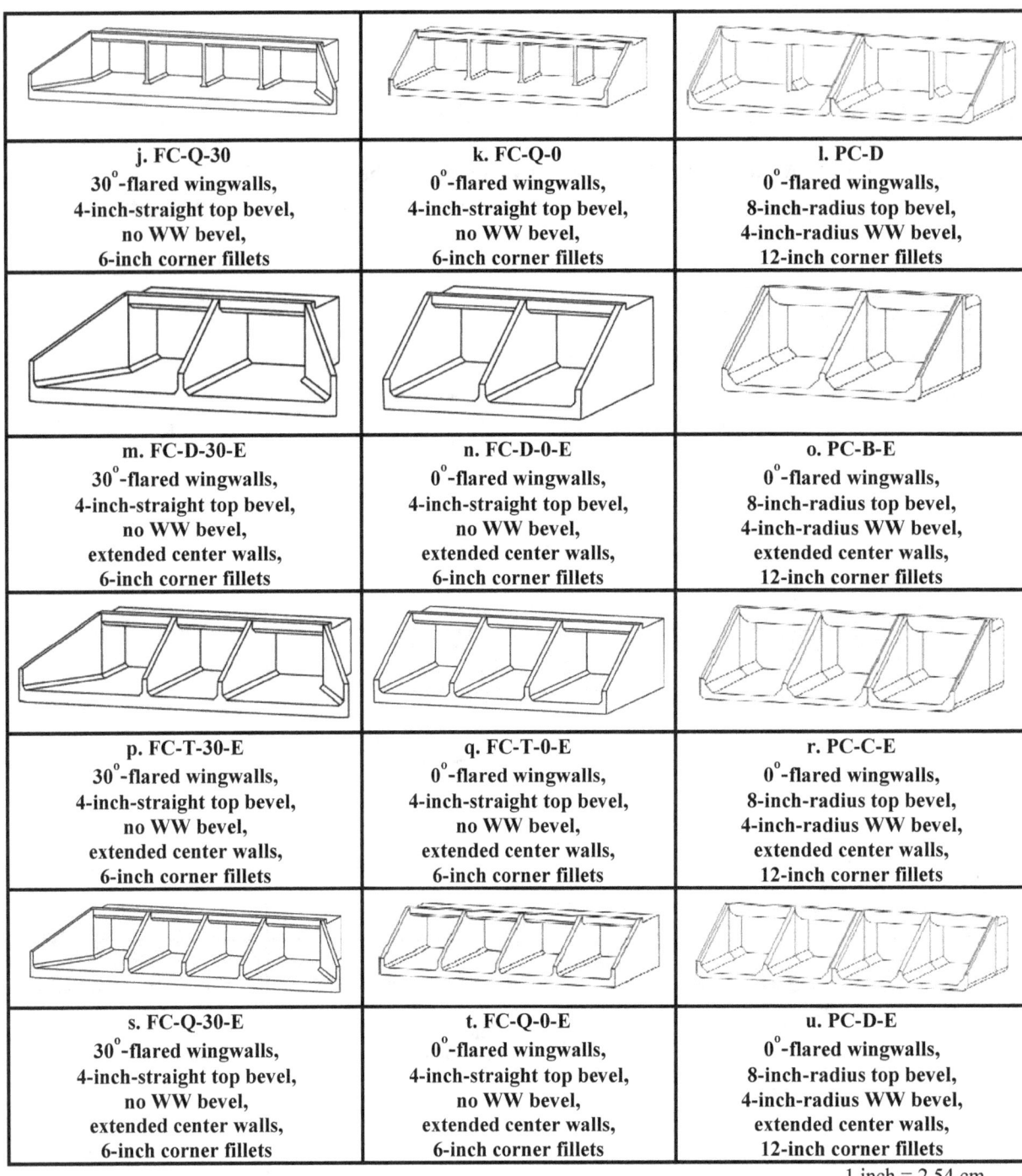

j. FC-Q-30 30°-flared wingwalls, 4-inch-straight top bevel, no WW bevel, 6-inch corner fillets	k. FC-Q-0 0°-flared wingwalls, 4-inch-straight top bevel, no WW bevel, 6-inch corner fillets	l. PC-D 0°-flared wingwalls, 8-inch-radius top bevel, 4-inch-radius WW bevel, 12-inch corner fillets
m. FC-D-30-E 30°-flared wingwalls, 4-inch-straight top bevel, no WW bevel, extended center walls, 6-inch corner fillets	n. FC-D-0-E 0°-flared wingwalls, 4-inch-straight top bevel, no WW bevel, extended center walls, 6-inch corner fillets	o. PC-B-E 0°-flared wingwalls, 8-inch-radius top bevel, 4-inch-radius WW bevel, extended center walls, 12-inch corner fillets
p. FC-T-30-E 30°-flared wingwalls, 4-inch-straight top bevel, no WW bevel, extended center walls, 6-inch corner fillets	q. FC-T-0-E 0°-flared wingwalls, 4-inch-straight top bevel, no WW bevel, extended center walls, 6-inch corner fillets	r. PC-C-E 0°-flared wingwalls, 8-inch-radius top bevel, 4-inch-radius WW bevel, extended center walls, 12-inch corner fillets
s. FC-Q-30-E 30°-flared wingwalls, 4-inch-straight top bevel, no WW bevel, extended center walls, 6-inch corner fillets	t. FC-Q-0-E 0°-flared wingwalls, 4-inch-straight top bevel, no WW bevel, extended center walls, 6-inch corner fillets	u. PC-D-E 0°-flared wingwalls, 8-inch-radius top bevel, 4-inch-radius WW bevel, extended center walls, 12-inch corner fillets

1 inch = 2.54 cm

Figure 49. Sketches. Models tested for effects of multiple barrels—*Continued*..

Multiple Barrels Versus Single Barrel

Figures 50 and 51 show that there is almost no difference in the performance of multiple barrels and a single barrel culvert for unsubmerged inlet control. For submerged inlet control, figures 50 and 51 show a slight hydraulic advantage for the multiple barrel field cast models when compared to the single barrel model. Figure 52 indicates a fairly substantial hydraulic advantage for the multiple barrel precast models when compared to the single barrel model for submerged inlet control. Several of the precast models were retested to verify the multiple barrel results.

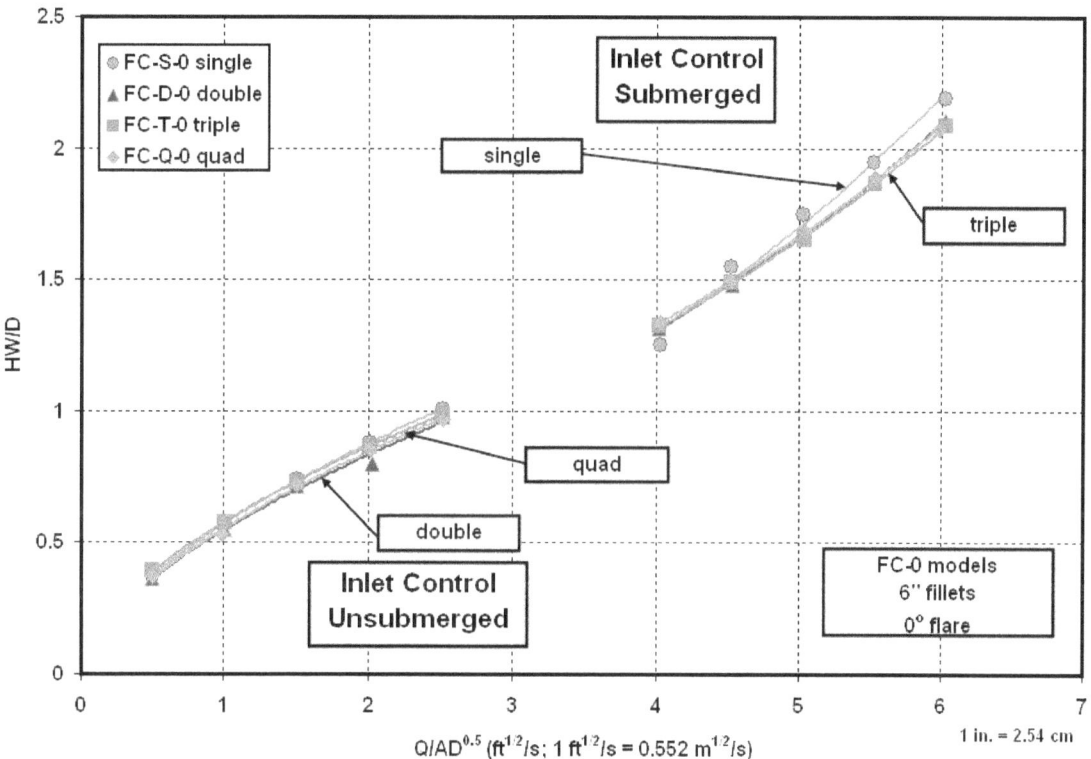

Figure 50. Graph. Inlet control comparison, field cast 0-degree-flared wingwall models.

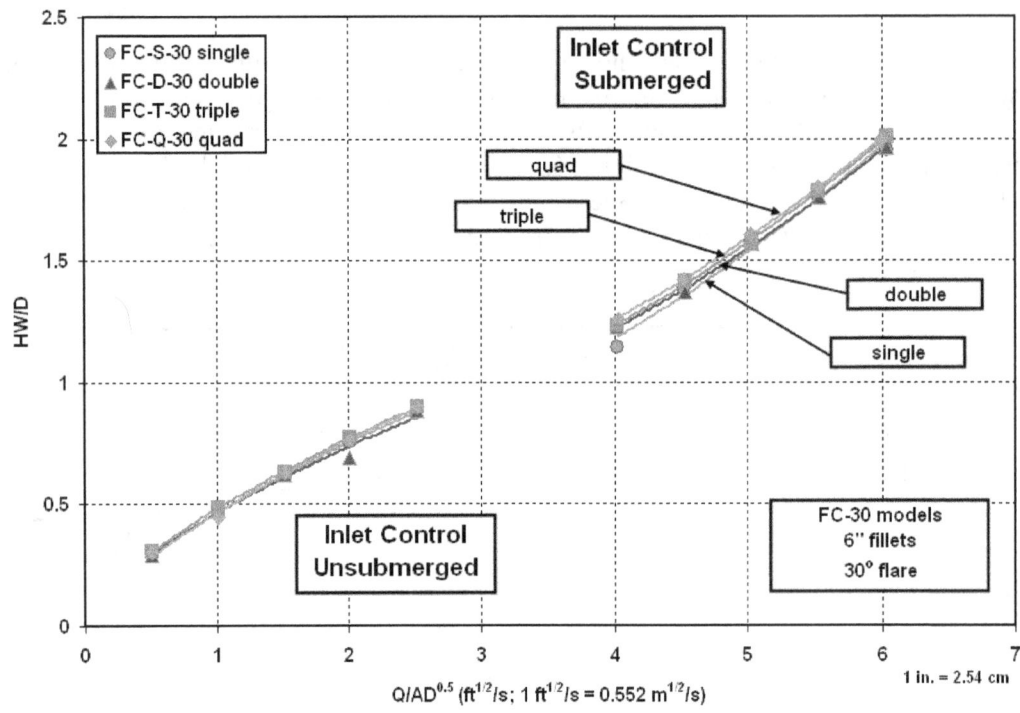

Figure 51. Graph. Inlet control comparison, field cast 30-degree-flared wingwall models.

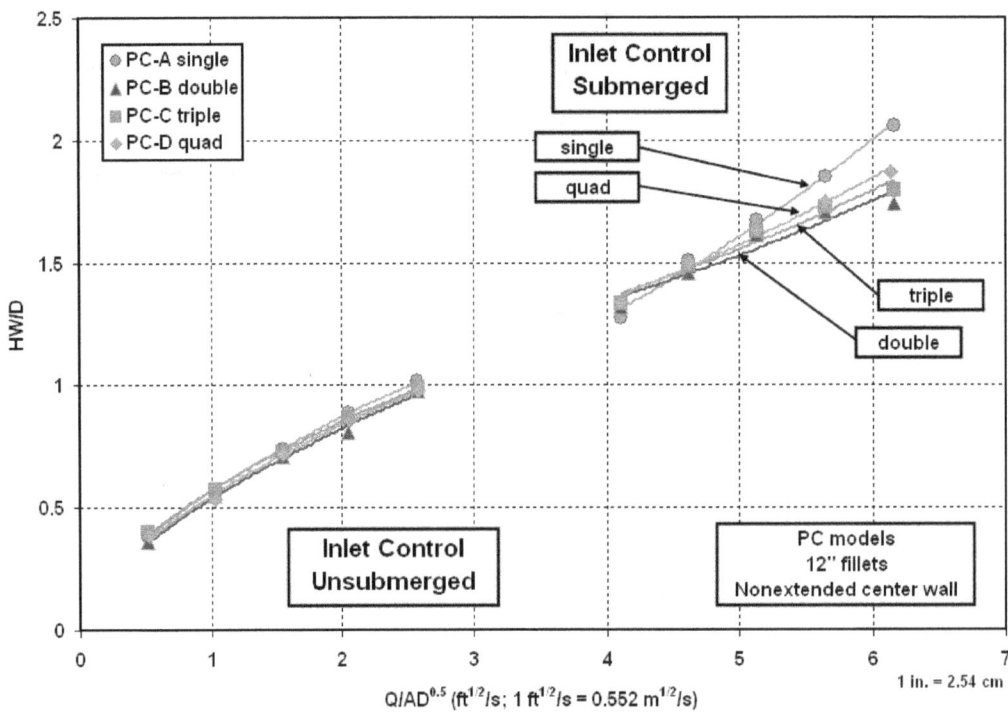

Figure 52. Graph. Inlet control comparison, precast models.

The coefficients derived for the multiple barrel tests (figures 50 to 58) are summarized in table 2. The outlet control results in table 2 were somewhat inconsistent. The K_e values for the field cast models with 0-degree wingwalls were almost identical for the single-barrel, double-barrel, triple-barrel, and quad-barrel models and averaged 0.52. These results support the practice of using the single barrel coefficients for multiple barrel analyses. For the 30-degree-flared wingwall models, the K_e values averaged 0.32 for the multiple barrel models when compared to K_e = 0.26 for the single barrel model. While not a big difference, it was counter to the inlet control indication in that the multiple barrels looked slightly worse than the single barrel for outlet control. For the precast models, the average K_e for the multiple barrel models was 0.54, and the K_e for the single barrel model was 0.33.

Although there are slight differences in single barrel and multiple barrel performance that were worthy of documentation, it is reasonable to combine the double-barrel, triple-barrel, and quad-barrel results.

Table 2. Summary of inlet and outlet control coefficients for models tested for effects of multiple barrels.

Model	Slope (percent)	Fillets (inches)	K_e	K Form 2 Eq.	M Form 2 Eq.	c	Y
FC-S-0	3	6		0.57	0.62	0.0448	0.56
	0.7	6	0.50				
FC-S-30	3	6		0.47	0.68	0.0394	0.53
	0.7	6	0.26				
FC-D-0	3	6		0.55	0.61	0.0391	0.66
	0.7	6	0.52				
FC-D-0-E	3	6		0.55	0.60	0.0394	0.65
	0.7	6	0.53				
FC-D-30	3	6		0.46	0.66	0.0366	0.61
	0.7	6	0.34				
FC-D-30-E	3	6		0.46	0.67	0.0381	0.59
	0.7	6	0.31				
FC-T-0	3	6		0.58	0.58	0.0377	0.69
	0.7	6	0.54				
FC-T-0-E	3	6		0.60	0.57	0.0405	0.67
	0.7	6	0.58				
FC-T-30	3	6		0.48	0.67	0.0379	0.60
	0.7	6	0.31				
FC-T-30-E	3	6		0.51	0.64	0.0405	0.60
	0.7	6	0.32				

Table 2. Summary of inlet and outlet control coefficients for models tested for effects of multiple barrels—*Continued.*

Model	Slope (percent)	Fillets (inches)	K_e	K Form 2 Eq.	M Form 2 Eq.	c	Y
FC-Q-0	3	6		0.55	0.61	0.0377	0.71
	0.7	6	0.52				
FC-Q-0-E	3	6		0.58	0.59	0.0418	0.64
	0.7	6	0.50				
FC-Q-30	3	6		0.47	0.71	0.0372	0.64
	0.7	6	0.32				
FC-Q-30-E	3	6		0.50	0.65	0.0398	0.60
	0.7	6	0.34				
PC-A	3	12		0.57	0.61	0.0361	0.68
	0.7	12	0.33				
PC-B	3	12		0.54	0.60	0.0253	0.90
	0.7	12	0.49				
PC-B-E	3	12		0.57	0.58	0.0315	0.81
	0.7	12	0.56				
PC-C	3	12		0.57	0.57	0.0219	0.98
	0.7	12	0.54				
PC-C-E	3	12		0.59	0.58	0.0263	0.91
	0.7	12	0.51				
PC-D	3	12		0.55	0.61	0.0250	0.92
	0.7	12	0.59				
PC-D-E	3	12		0.60	0.54	0.0296	0.85
	0.7	12	0.58				

1 inch = 2.54 cm

Notes: For empty cells, data are not available or not applicable. Eq. is equation. The form 2 equation is identified in chapter 3.

Effect of Wingwall-Flare Angle

As the number of barrels increases, a smaller percentage of flow is influenced by the wingwalls, with less advantage to having flared wingwalls. To visualize the effect, observe the space (Δ) between the 30-degree- and the 0-degree-flared wingwall performance curves in figures 53–56. Although the gap never closes, the curves become closer as the number of barrels increases. Interestingly, the PC multiple barrel models with the optimum curved top plates outperformed the 30-degree-flared wingwall multiple barrel models at headwater to culvert depth ratios greater than

1.5. It is reasonable to expect that the optimum top plate bevel will have a more pronounced effect on performance at the high headwater depths as the number of barrels and total span increase.

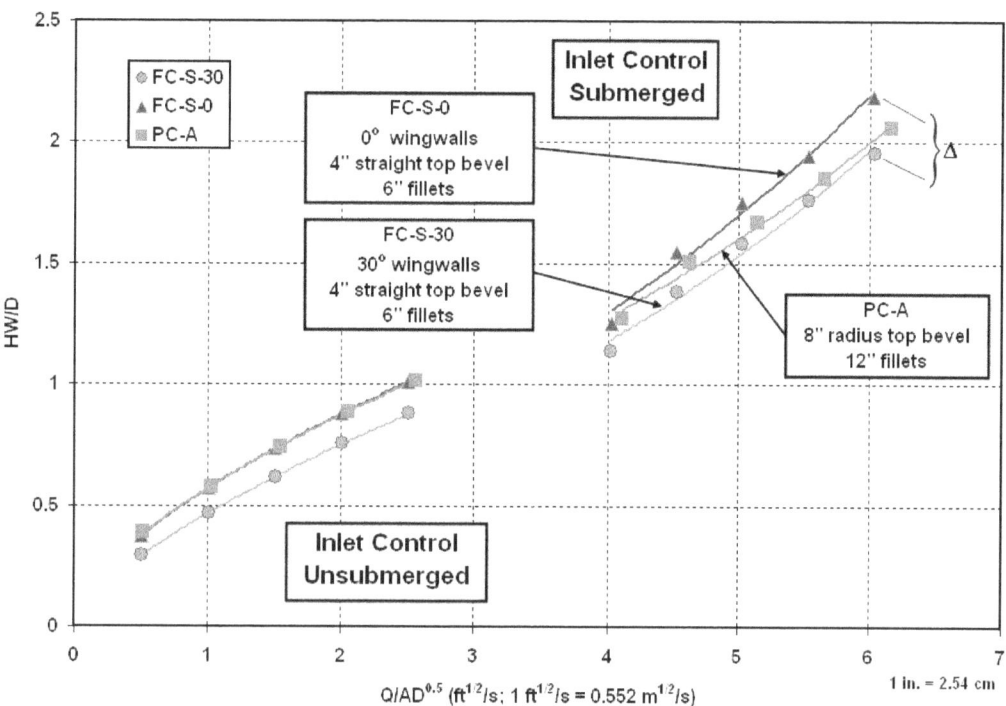

Figure 53. Graph. Inlet control comparison, single-barrel models.

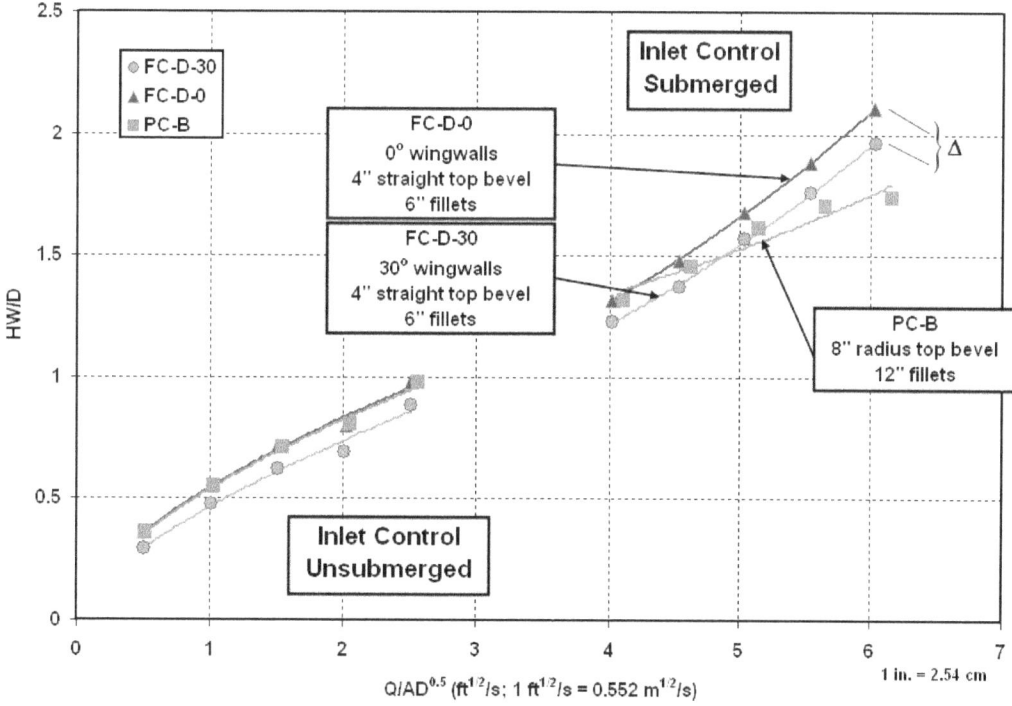

Figure 54. Graph. Inlet control comparison, double-barrel models.

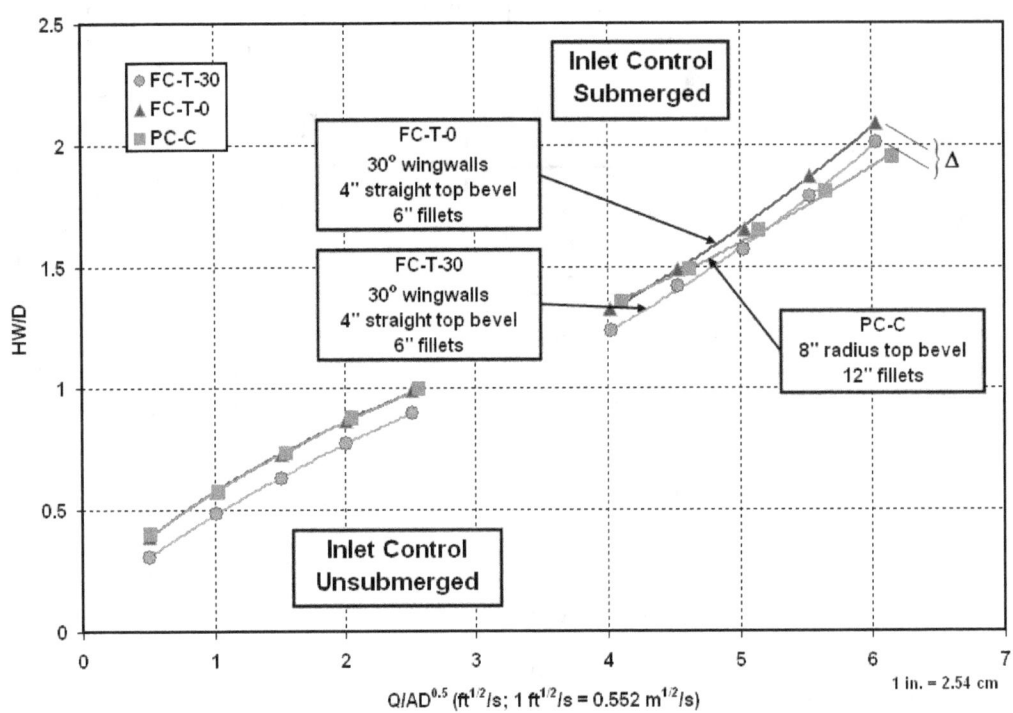

Figure 55. Graph. Inlet control comparison, triple-barrel models.

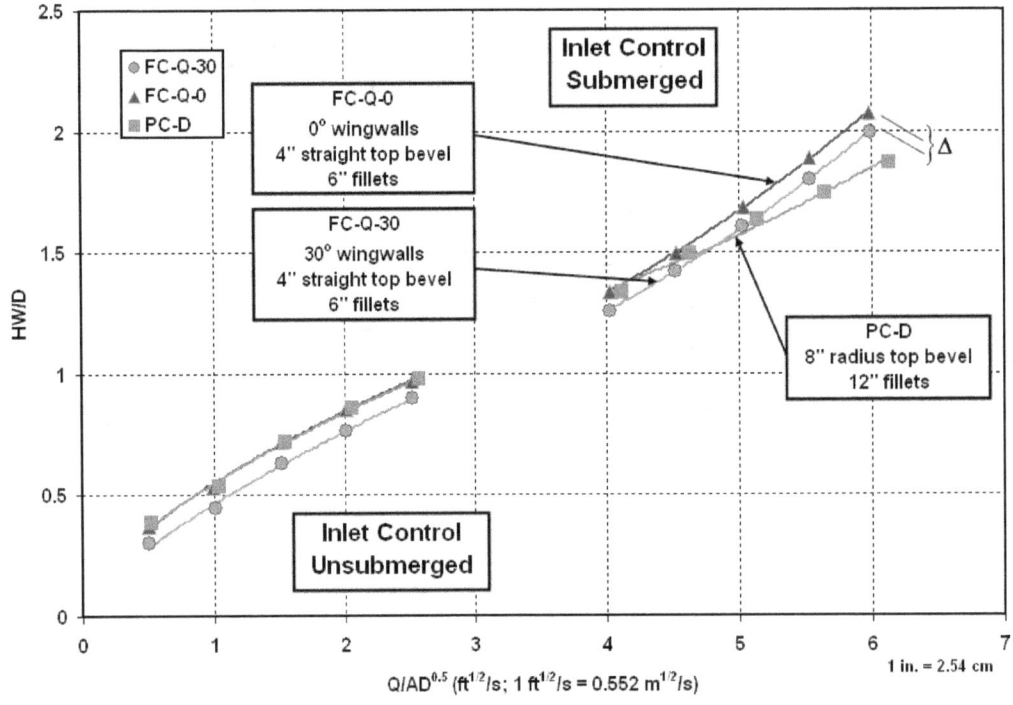

Figure 56. Graph. Inlet control comparison, quadruple-barrel models.

Effect of Center Wall Extension

There is no hydraulic advantage or disadvantage to extending center walls for multiple barrel culverts. Figure 57 is typical of all results for the field cast model in that, for the inlet control tests, the performance for the extended center walls matched the performance for the nonextended center walls. Field cast inlets show nearly the same performance with or without center wall extensions. Figure 58 shows a similar comparison for the precast models with and without extended center walls. The data point for the highest discharge intensity for the PC-B model was omitted from the trend line in figure 58 because it was considered an outlier point.

For outlet control, the K_e values tabulated in table 2 are almost identical in every case for the extended center walls and the corresponding nonextended center walls, including the case of precast models.

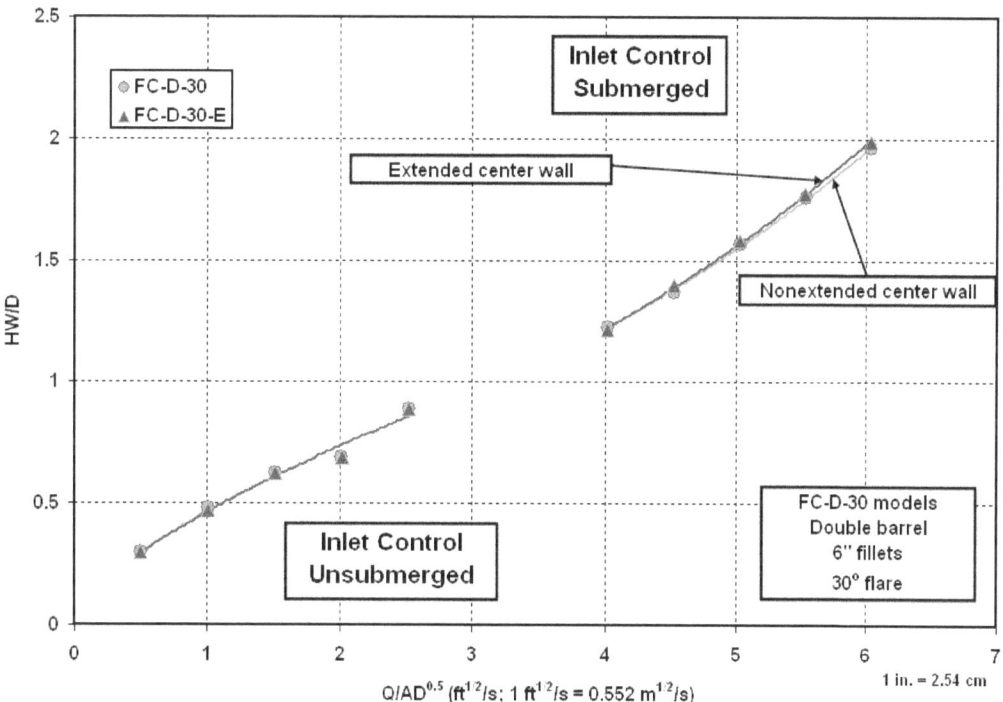

Figure 57. Graph. Inlet control comparison, extended or nonextended center walls, field cast model.

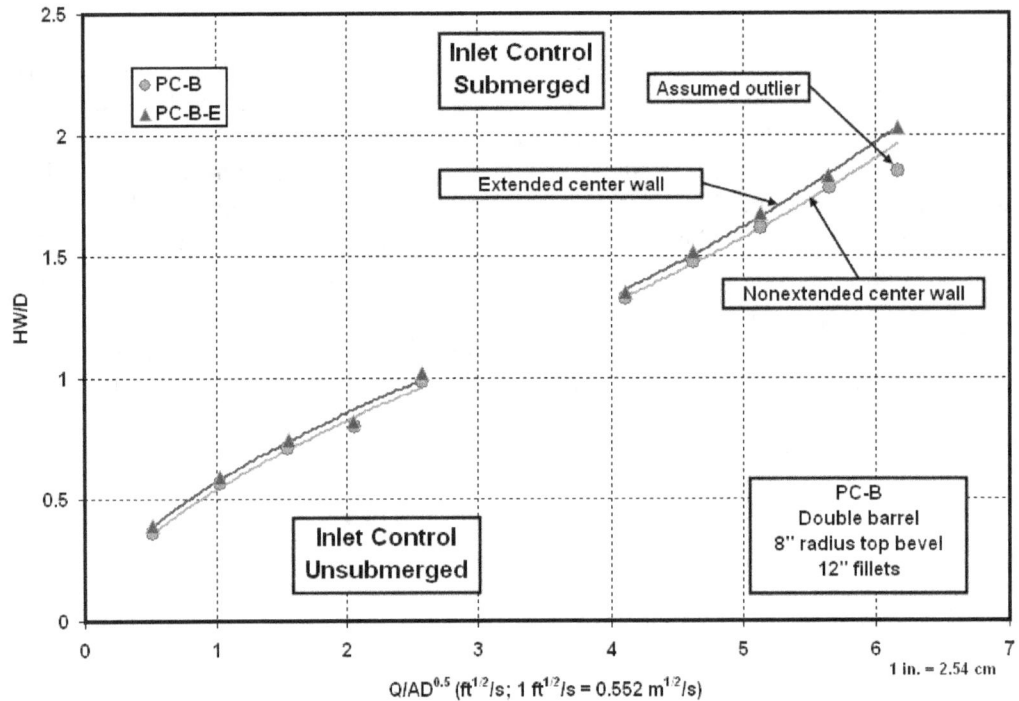

Figure 58. Inlet control comparison, extended or nonextended center walls, precast model.

EFFECTS OF SPAN-TO-RISE RATIO

Models tested concerning the span-to-rise ratio are illustrated in figure 59.

Multiple Span-to-Rise Versus Basic 1:1 Span-to-Rise

For inlet control in field cast models with 0-degree wingwalls (FC-S-0), figure 60 shows that a very slight loss in performance might occur as the span-to-rise ratio increases for unsubmerged flow but almost no effect for submerged flow. Figure 61 shows similar results for the precast models. Figure 62 shows a discernable decrease in performance in the submerged flow zone as the span-to-rise ratio increases for the field cast models with 30-degree-flared wingwalls (FC-S-30). This decrease in performance can be attributed in part to the diminishing effects of the wingwall flare angle with increasing span-to-rise, as discussed shortly. Table 3 at the end of this section also contains data on inlet control.

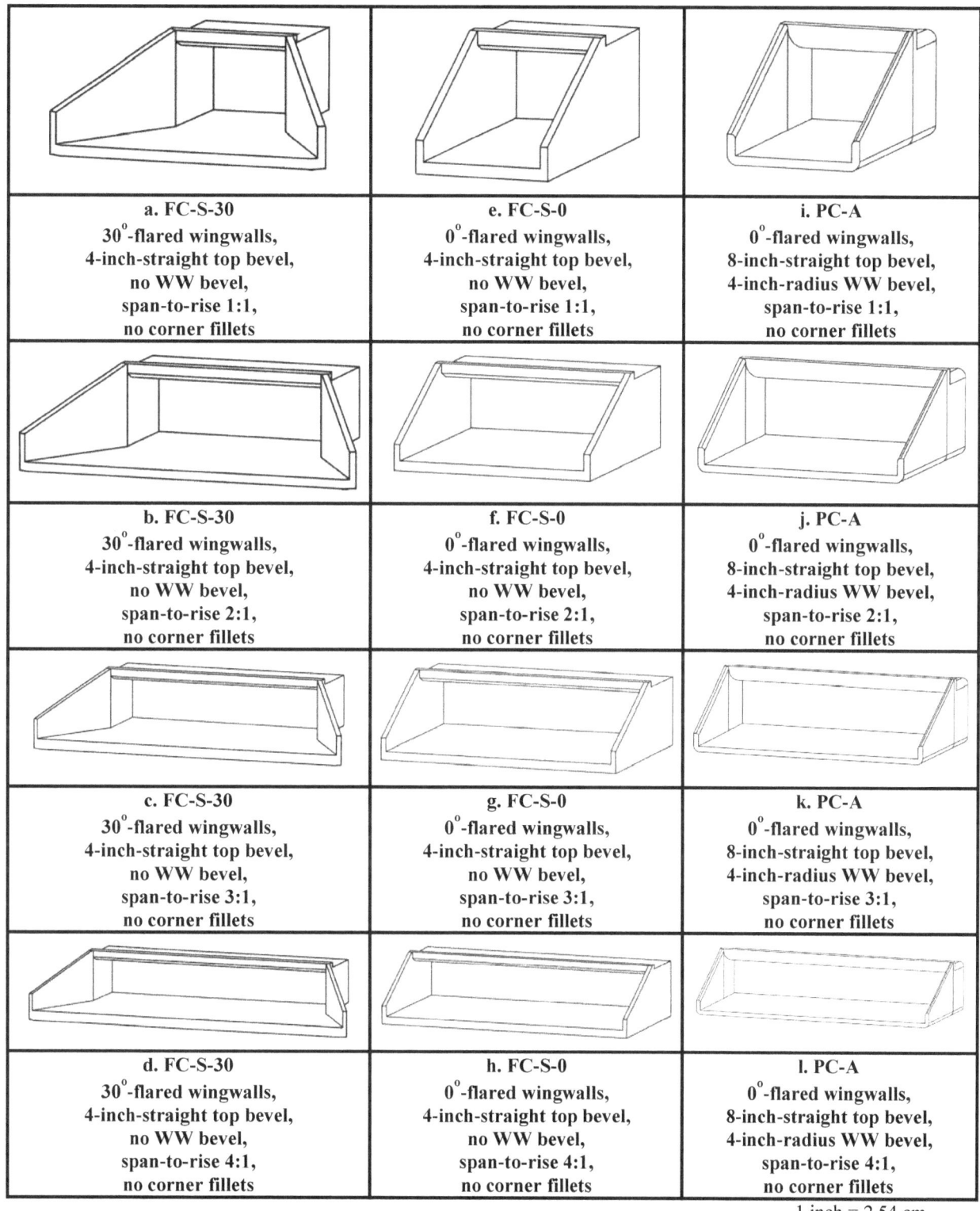

a. FC-S-30 30°-flared wingwalls, 4-inch-straight top bevel, no WW bevel, span-to-rise 1:1, no corner fillets	**e. FC-S-0** 0°-flared wingwalls, 4-inch-straight top bevel, no WW bevel, span-to-rise 1:1, no corner fillets	**i. PC-A** 0°-flared wingwalls, 8-inch-straight top bevel, 4-inch-radius WW bevel, span-to-rise 1:1, no corner fillets
b. FC-S-30 30°-flared wingwalls, 4-inch-straight top bevel, no WW bevel, span-to-rise 2:1, no corner fillets	**f. FC-S-0** 0°-flared wingwalls, 4-inch-straight top bevel, no WW bevel, span-to-rise 2:1, no corner fillets	**j. PC-A** 0°-flared wingwalls, 8-inch-straight top bevel, 4-inch-radius WW bevel, span-to-rise 2:1, no corner fillets
c. FC-S-30 30°-flared wingwalls, 4-inch-straight top bevel, no WW bevel, span-to-rise 3:1, no corner fillets	**g. FC-S-0** 0°-flared wingwalls, 4-inch-straight top bevel, no WW bevel, span-to-rise 3:1, no corner fillets	**k. PC-A** 0°-flared wingwalls, 8-inch-straight top bevel, 4-inch-radius WW bevel, span-to-rise 3:1, no corner fillets
d. FC-S-30 30°-flared wingwalls, 4-inch-straight top bevel, no WW bevel, span-to-rise 4:1, no corner fillets	**h. FC-S-0** 0°-flared wingwalls, 4-inch-straight top bevel, no WW bevel, span-to-rise 4:1, no corner fillets	**l. PC-A** 0°-flared wingwalls, 8-inch-straight top bevel, 4-inch-radius WW bevel, span-to-rise 4:1, no corner fillets

1 inch = 2.54 cm

Figure 59. Sketches. Models tested for effects of span-to-rise ratio.

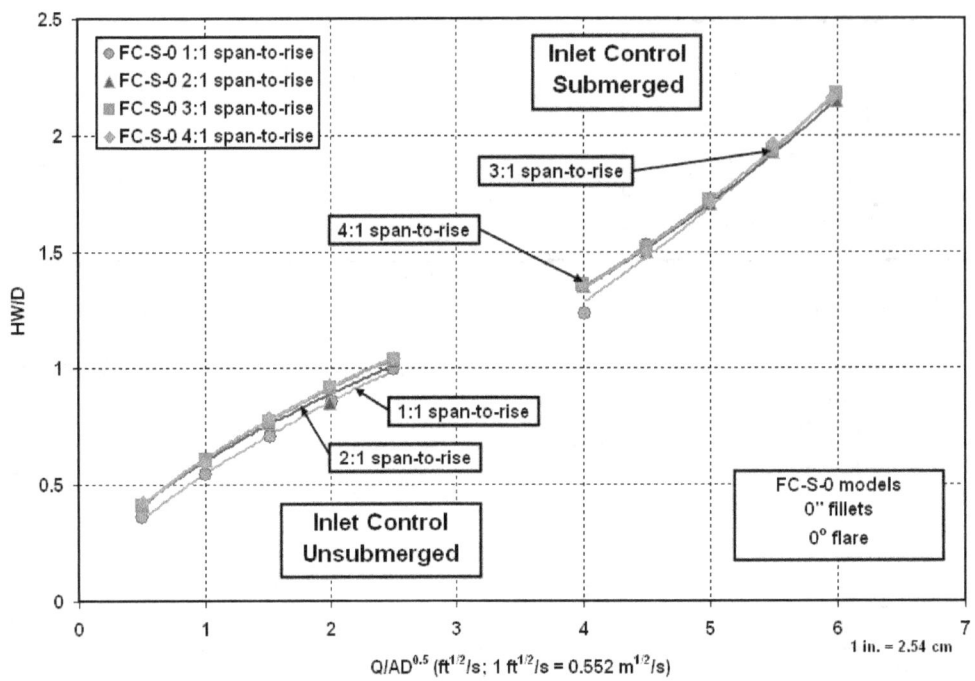

Figure 60. Graph. Inlet control comparison, FC-S-0 span-to-rise ratios.

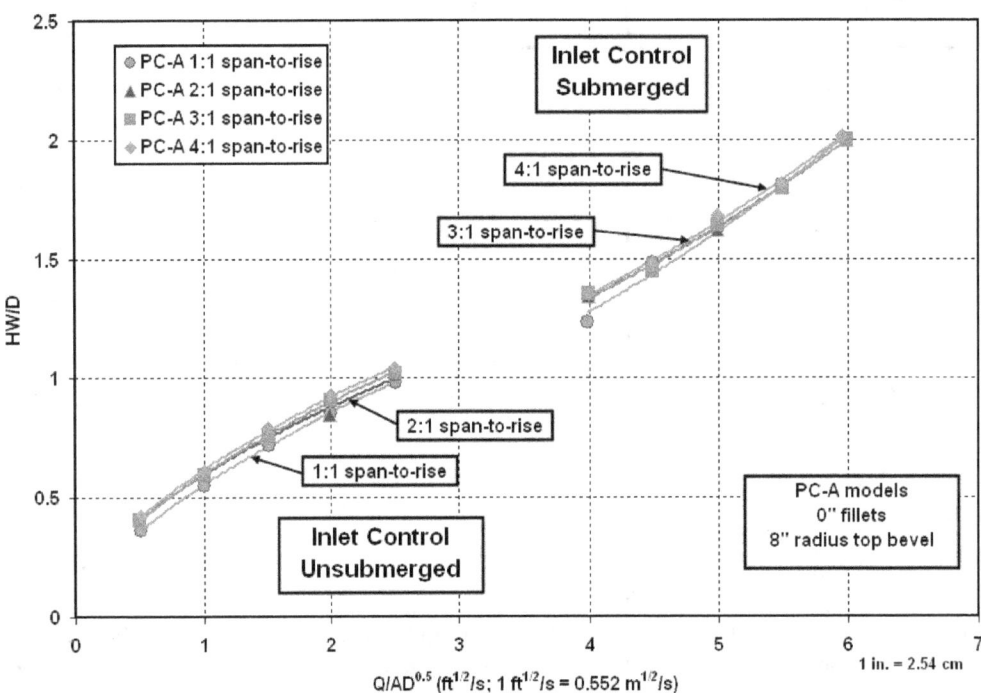

Figure 61. Graph. Inlet control comparison, PC-A span-to-rise ratios.

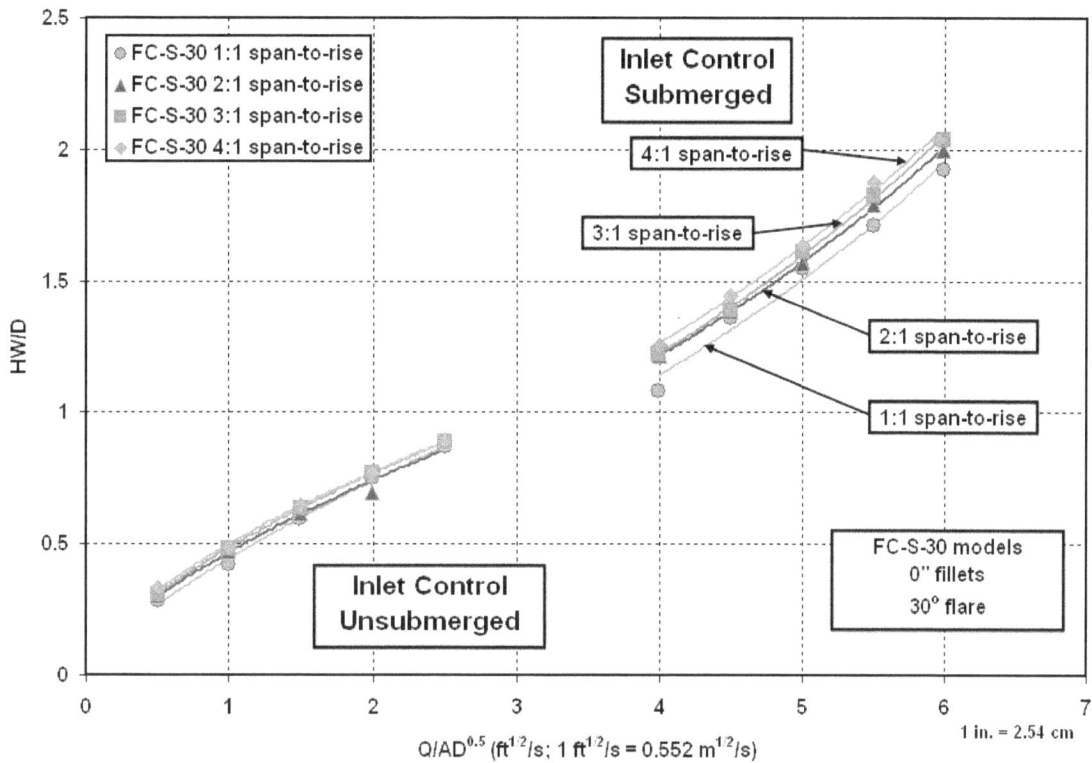

Figure 62. Graph. Inlet control comparison, FC-S-30 span-to-rise ratios.

For outlet control, the entrance loss coefficient K_e for the FC-S-0 models varied from 0.32 for the 3:1 span-to-rise ratio to 0.46 for the 1:1 ratio (table 3 at the end of this section), but there was no clear trend, and the variation can be attributed to experimental scatter. Similarly, for the PC-A models, the entrance loss coefficient K_e varied from 0.26 for the 4:1 span-to-rise ratio to 0.34 for the 2:1 ratio; again, there was no clear trend. There was a slight trend in the entrance loss coefficients for the FC-S-30 models in that K_e was 0.27 for the 1:1 and 2:1 span-to-rise ratios and was 0.19 and 0.18 for the 3:1 and 4:1 ratios, respectively. The trend was just the opposite from the effect observed for the inlet control tests. One may draw from the outlet control results the reasonable conclusions that the variations represent experimental scatter and that the basic 1:1 model entrance loss coefficients can be applied for various span-to-rise ratios.

Wingwall Flare and the Span-to-Rise Ratio

Figures 63–66 show that the wingwall flare angle has a diminishing effect as the span-to-rise ratio increases, which is analogous to the multiple barrel phenomenon. This observation can be visualized by inspecting the spacing, labeled as (Δ) in these figures, between the 0-degree-flared wingwall curves and the 30-degree-flared wingwall curves. As the span-to-rise and headwater depth ratios increase, the effect of the top plate bevel increases and the effect of the wingwall flare angle decreases, which explains why the PC models slightly outperformed the 30-degree-flared wingwall models for a few situations. This was similar to, but less pronounced than in, the case of the multiple barrel comparison.

The outlet control entrance loss coefficients listed in table 3 for the FC-S-30 models did not, however, support the assumption that the wingwall flare angle has a diminishing effect as the span-to-rise ratio increases. The outlet control loss coefficient K_e was 0.27 for the basic 1:1 span-to-rise ratio, but decreased to 0.18 for the 4:1 ratio.

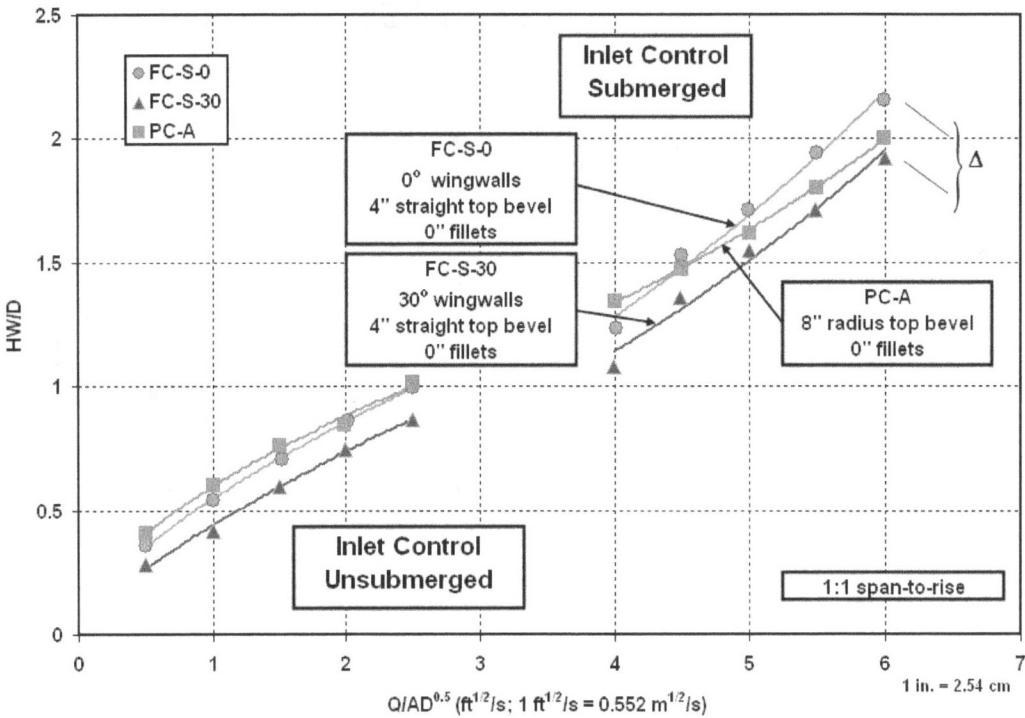

Figure 63. Graph. Inlet control comparison, 1:1 span-to-rise ratio.

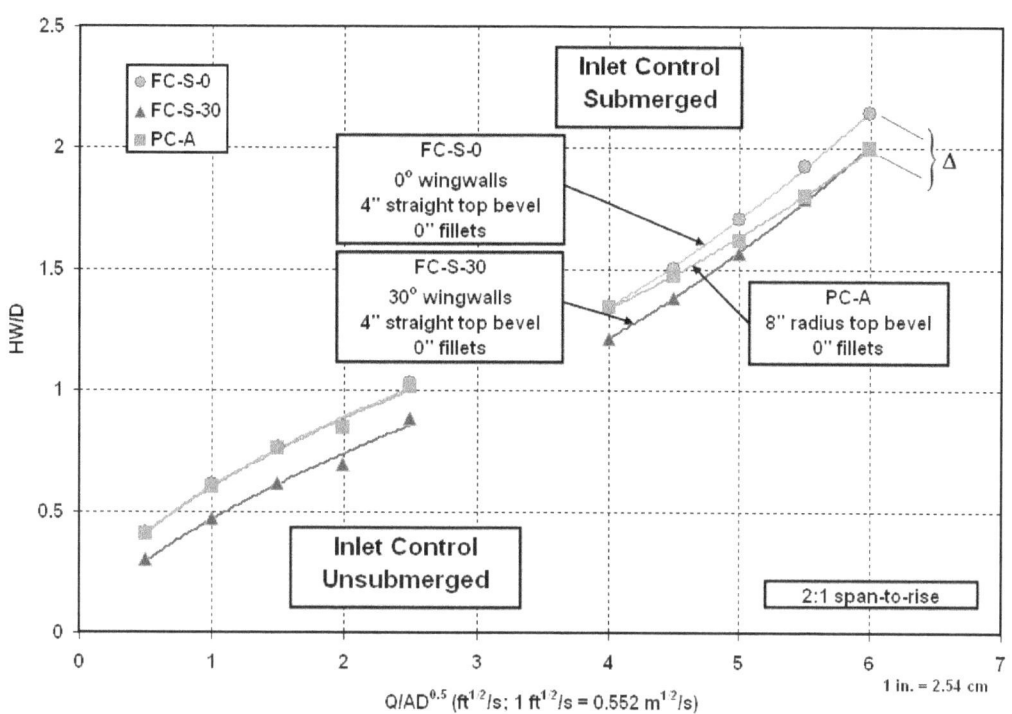

Figure 64. Graph. Inlet control comparison, 2:1 span-to-rise ratio.

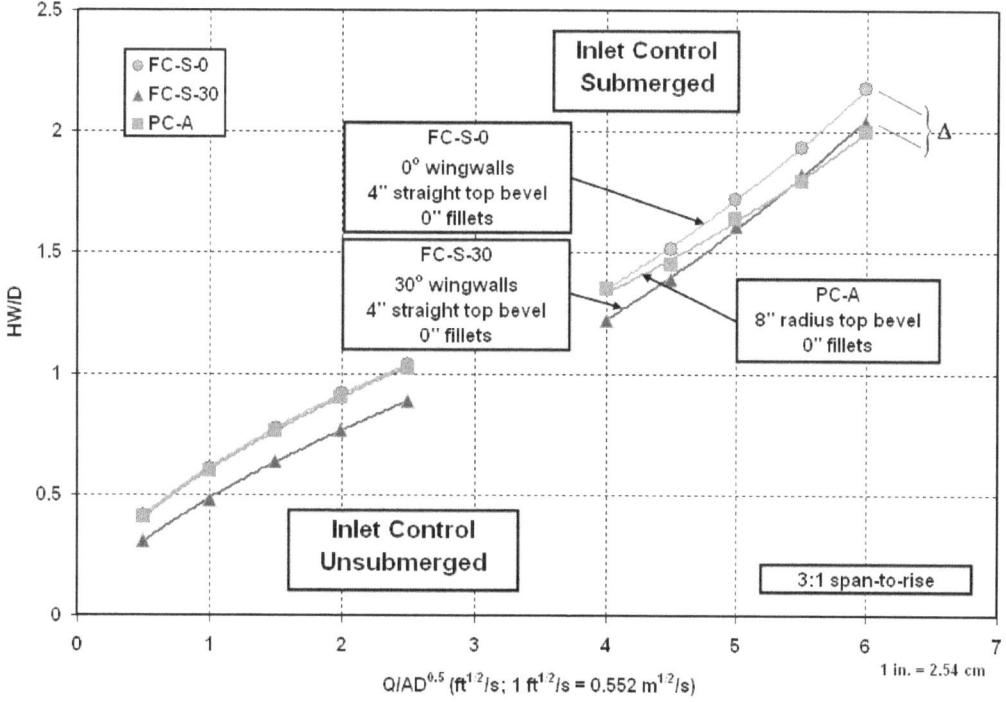

Figure 65. Graph. Inlet control comparison, 3:1 span-to-rise ratio.

57

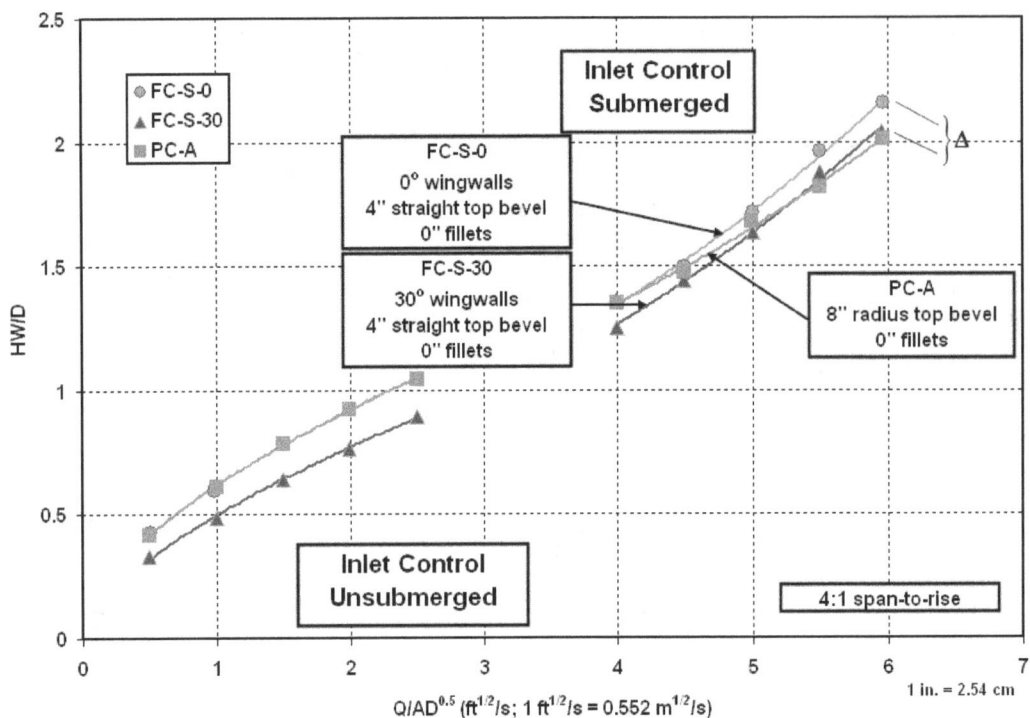

Figure 66. Graph. Inlet control comparison, 4:1 span-to-rise ratio.

Table 3. Summary of inlet and outlet control coefficients for models tested for effects of span-to-rise ratio.

Model	Span:Rise	Slope (percent)	Fillets (inches)	K_e	K Form 2 Eq.	M Form 2 Eq.	c	Y
FC-S-0	1:1	3	0		0.55	0.64	0.0453	0.54
		0.7	0	0.46				
	2:1	3	0		0.60	0.56	0.0404	0.68
		0.7	0	0.40				
	3:1	3	0		0.61	0.58	0.0413	0.67
		0.7	0	0.32				
	4:1	3	0		0.62	0.57	0.0421	0.65
		0.7	0	0.40				
FC-S-30	1:1	3	0		0.44	0.74	0.0403	0.48
		0.7	0	0.27				
	2:1	3	0		0.47	0.66	0.0397	0.56
		0.7	0	0.22				
	3:1	3	0		0.48	0.66	0.0414	0.54
		0.7	0	0.19				
	4:1	3	0		0.50	0.63	0.0410	0.59
		0.7	0	0.18				
PC-A	1:1	3	0		0.56	0.62	0.0371	0.66
		0.7	0	0.27				
	2:1	3	0		0.60	0.56	0.0329	0.79
		0.7	0	0.34				
	3:1	3	0		0.60	0.58	0.0331	0.79
		0.7	0	0.29				
	4:1	3	0		0.62	0.57	0.0340	0.79
		0.7	0	0.26				

1 inch = 2.54 cm

Notes: For empty cells, data are not available or not applicable. Eq. is equation. The form 2 equation is identified in chapter 3.

EFFECTS OF HEADWALL SKEW

Skewed headwalls as illustrated in figure 67 were tested for skew angles of 0, 15, 30, and 45 degrees. Although it is not uncommon to see culvert installations with the approach flow actually skewed to the culvert alignment, this condition is considered bad practice and was not included in the test matrix.

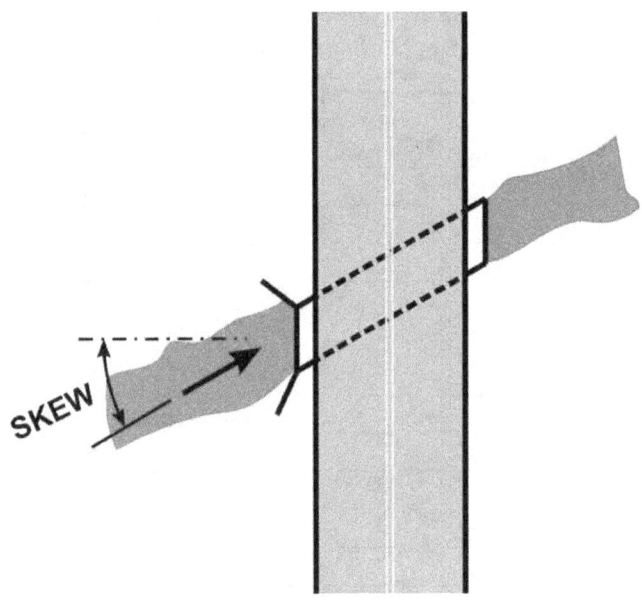

Figure 67. Sketch. Definition sketch for skew tests.

The field cast triple-barrel 30-degree-flared wingwalls model (FC-T-30) and the field cast single-barrel 30-degree-flared wingwalls model with a 3:1 span-to-rise ratio (FC-S-30 (3:1)) illustrated in figure 68 were used for these tests.

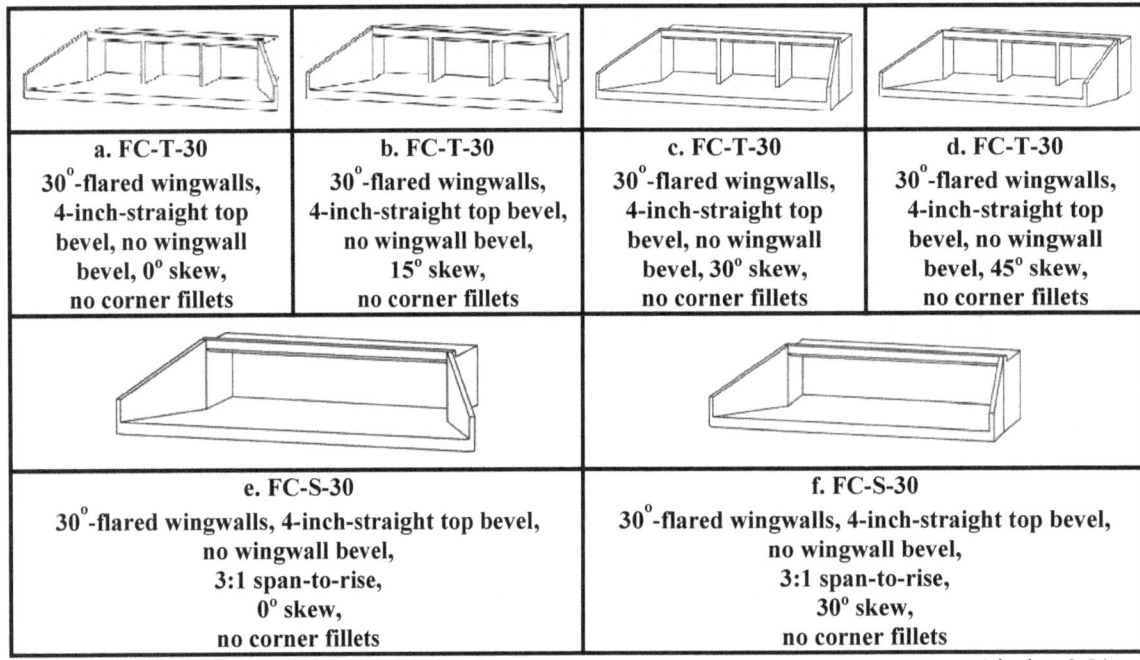

a. FC-T-30 30°-flared wingwalls, 4-inch-straight top bevel, no wingwall bevel, 0° skew, no corner fillets	**b. FC-T-30** 30°-flared wingwalls, 4-inch-straight top bevel, no wingwall bevel, 15° skew, no corner fillets	**c. FC-T-30** 30°-flared wingwalls, 4-inch-straight top bevel, no wingwall bevel, 30° skew, no corner fillets	**d. FC-T-30** 30°-flared wingwalls, 4-inch-straight top bevel, no wingwall bevel, 45° skew, no corner fillets

e. FC-S-30 30°-flared wingwalls, 4-inch-straight top bevel, no wingwall bevel, 3:1 span-to-rise, 0° skew, no corner fillets	**f. FC-S-30** 30°-flared wingwalls, 4-inch-straight top bevel, no wingwall bevel, 3:1 span-to-rise, 30° skew, no corner fillets

1 inch = 2.54 cm

Figure 68. Sketches. Models tested for effects of headwall skew.

Figure 69 is the plan view of the skewed inlets tested.

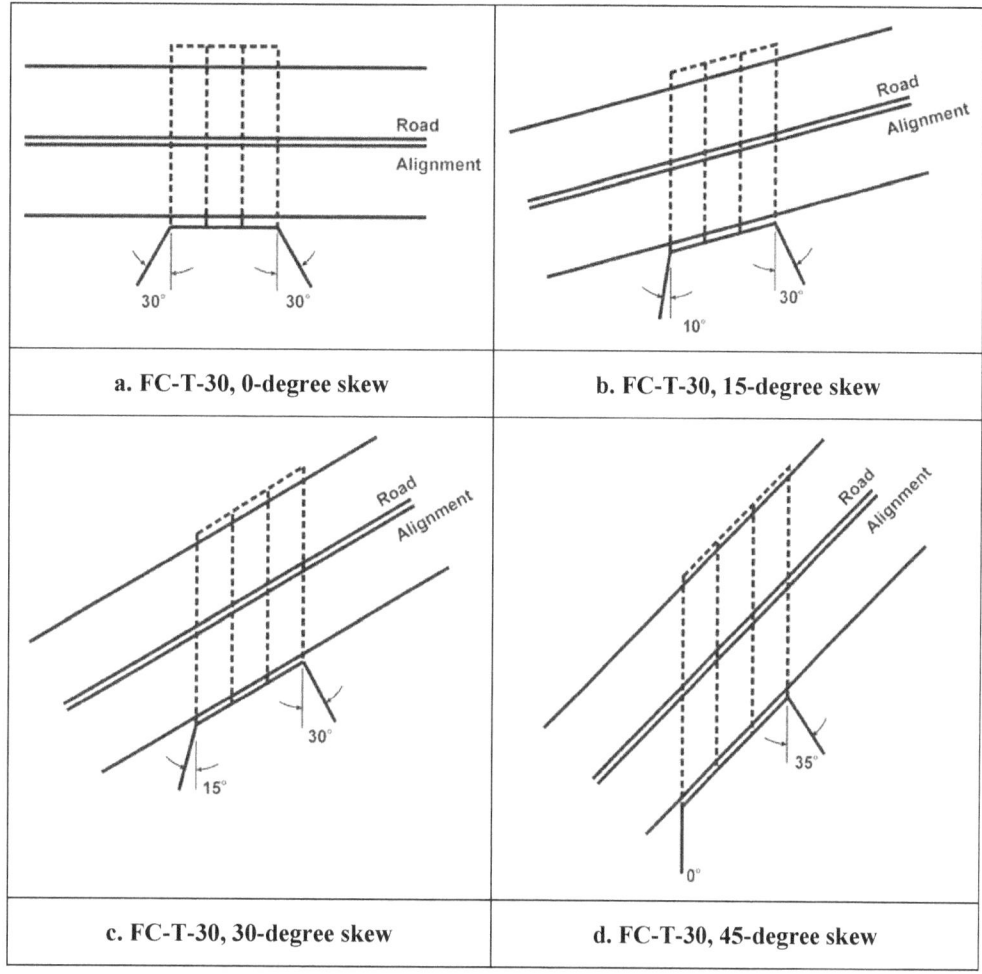

a. FC-T-30, 0-degree skew

b. FC-T-30, 15-degree skew

c. FC-T-30, 30-degree skew

d. FC-T-30, 45-degree skew

Figure 69. Diagrams. Plan view of skewed headwall models tested.

For inlet control, figure 70 shows that the performance curves for the three skewed headwalls (at 15, 30, and 45 degrees) do cluster but are separated from the performance curve for the 0-degree headwall. Consequently, it is reasonable to combine the 15-degree, 30-degree, and 45-degree skew curves for inlet control. The performance curve for the HDS-5, chart 12, scale 3, inlet, which was noted in that publication to be a good approximation for skews from 15 degrees to 45 degrees, is plotted in this figure for comparison, but it does not compare favorably with the skewed headwall models tested in this study. HDS-5 does not specify if that inlet represents a skewed headwall or a skewed flow alignment.

The outlet control entrance loss coefficient K_e could be averaged at 0.36 for 0-degree and 15-degree skew angles, but increased to 0.46 for 30-degree and 45-degree skew angles. Table 4 summarizes the inlet and outlet control coefficients.

61

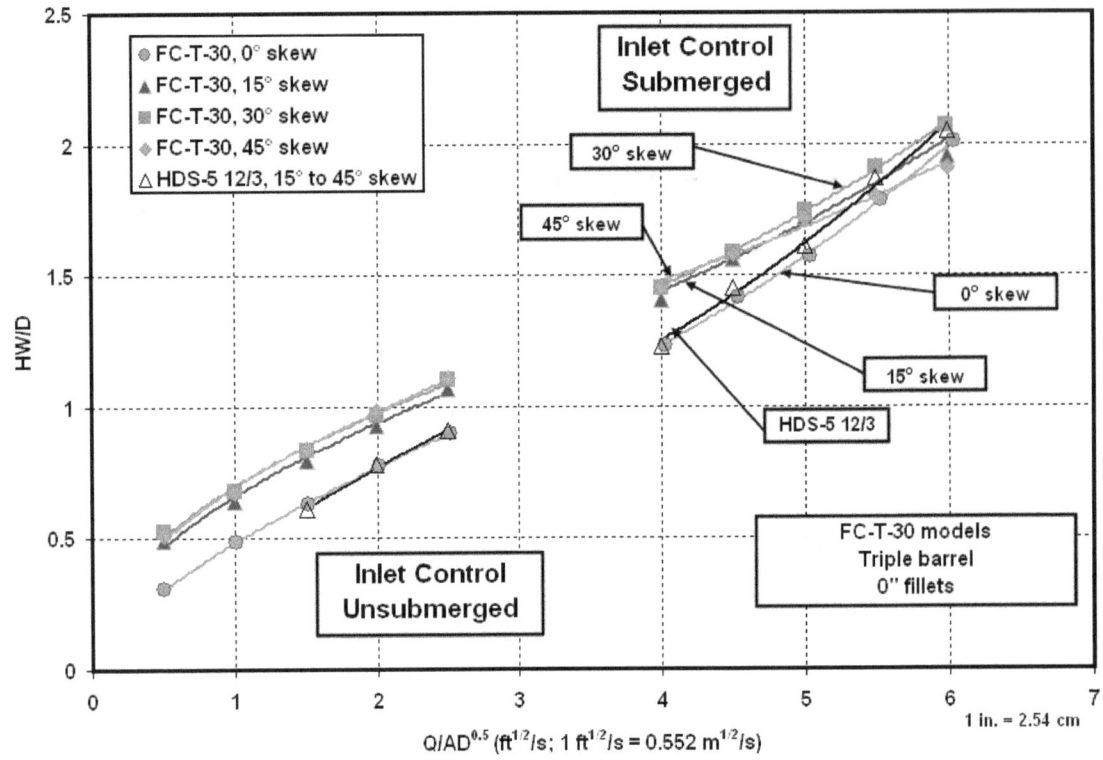

Figure 70. Inlet control comparison, skew angles.

Table 4. Summary of inlet and outlet control coefficients for models tested for effects of headwall skew.

Model	Span: Rise	Skew	Slope (percent)	Fillets (inches)	K_e	K Form 2 Eq.	M Form 2 Eq.	c	Y
FC-S-30	3:1	0°	3	0		0.48	0.66	0.0414	0.54
			0.7	0	0.19				
		30°	3	0		0.68	0.46	0.0306	0.89
			0.7	0	0.39				
FC-T-30	3:1	0°	3	0		0.48	0.67	0.0369	0.62
			0.7	0	0.35				
		15°	3	0		0.66	0.50	0.0289	0.95
			0.7	0	0.36				
		30°	3	0		0.70	0.48	0.0312	0.94
			0.7	0	0.44				
		45°	3	0		0.69	0.50	0.0224	1.10
			0.7	0	0.46				

1 inch = 2.54 cm

Notes: For empty cells, data are not available or not applicable. Eq. is equation. The form 2 equation is identified in chapter 3.

OUTLET CONTROL ENTRANCE LOSS COEFFICIENTS K_e FOR LOW FLOWS (UNSUBMERGED CONDITIONS)

In culvert design, outlet control is typically associated with full or nearly full barrel flow. Outlet control can occur, however, for partly full flow if the culvert barrel slope is flat enough. This condition is often encountered when environmental considerations, such as allowing the passage of fish, mandate a flatter slope. Outlet control loss coefficients are usually tabulated as constants for inlet types. Whereas some experimental scatter is expected, the scatter in the loss coefficients for unsubmerged flow in this study was so extreme that researchers decided to separate the unsubmerged data from the submerged data for the outlet control experiments. The unsubmerged loss coefficients are not considered reliable for implementation, but the results and attempted analyses are useful for future research.

Table 5 summarizes the average entrance loss coefficients for unsubmerged flow conditions. Not all the data points could be used to derive the average coefficients because some of the values were so far out of range that they were discounted during the data reduction phase of the study. The worst values seemed to occur at the lowest flows, and the problem was attributed to the resolution limits of the pressure sensors. The accuracy of the current pressure transducer used is ± 5 mm (0.196 inch). At the very low flows, the actual head losses measured were less than the resolution of the pressure sensor. These measurements, which were mostly "noise" in the sensors, were divided by a velocity head, which was near zero, to compute the entrance loss coefficients, which were often an order of magnitude larger than expected. In general, the average entrance loss coefficients for unsubmerged conditions tended to be higher than the coefficients for submerged conditions. It was difficult, however, to determine if this tendency was real or just a result of the selection process on which values to include in the average.

Effects of Reynolds Number on K_e

One hypothesis considered as a possible explanation for the large scatter in the entrance loss coefficient at low flows was that K_e depended on the Reynolds number. If that hypothesis were true, there should be a correlation between K_e and the Reynolds number regardless of whether the culvert were submerged.

The Reynolds number hypothesis was tested using full barrel flow with the HDS-5, chart 8, scale 3, inlet because it was easier to control Reynolds numbers for full barrel flow than for free surface flow. Data was acquired with the Reynolds number held constant while the headwater depth ratio (HW/D) was varied. This process was repeated for a range of Reynolds numbers up to the maximum Reynolds number that could be accomplished with the experimental apparatus.

The Reynolds number is defined as a characteristic velocity times a characteristic length divided by the kinematic viscosity. The hydraulic radius times four was used as the characteristic diameter. Only submerged conditions were tested, so the characteristic length was the barrel diameter, D. The average flow velocity was calculated by dividing the discharge by the area of the barrel. The entrance loss coefficients for the HDS-5, chart 8, scale 3, inlet for Reynolds numbers ranging from 65000 to 260000 are presented in figure 71.

Table 5. Summary of outlet control entrance loss coefficients, K$_e$, for low flows.

Model	Slope (percent)	Fillets (inches)	Number of Barrels	Span:Rise	Skew	Unsubmerged K$_e$ *
FC-S-0	0.7	0	1	1:1	0°	0.73
FC-S-0	0.7	6	1	1:1	0°	0.90
FC-S-0	0.7	12	1	1:1	0°	0.9
PC-A	0.7	0	1	1:1	0°	0.67
PC-A	0.7	6	1	1:1	0°	0.63
PC-A	0.7	12	1	1:1	0°	0.56
FC-S-30	0.7	0	1	1:1	0°	0.39
FC-S-30	0.7	6	1	1:1	0°	0.71
FC-S-0	0.7	6	1	1:1	0°	0.9
FC-S-30	0.7	6	1	1:1	0°	0.71
FC-D-0	0.7	6	2	1:1	0°	0.71
FC-D-0-E	0.7	6	2	1:1	0°	0.32
FC-D-30	0.7	6	2	1:1	0°	0.74
FC-D-30-E	0.7	6	2	1:1	0°	0.42
FC-T-0	0.7	6	3	1:1	0°	0.6
FC-T-0-E	0.7	6	3	1:1	0°	0.8
FC-T-30	0.7	6	3	1:1	0°	0.48
FC-T-30-E	0.7	6	3	1:1	0°	0.41
FC-Q-0	0.7	6	4	1:1	0°	0.83
FC-Q-0-E	3.0	6	4	1:1	0°	0.87
FC-Q-30	0.7	6	4	1:1	0°	0.38
FC-Q-30-E	0.7	6	4	1:1	0°	0.38
PC-A	0.7	12	1	1:1	0°	0.56
PC-B	0.7	12	2	1:1	0°	0.96
PC-B-E	0.7	12	2	1:1	0°	0.75
PC-C	0.7	12	3	1:1	0°	0.94
PC-C-E	0.7	12	3	1:1	0°	0.96
PC-D	0.7	12	4	1:1	0°	0.91
PC-D-E	0.7	12	4	1:1	0°	0.93
FC-S-0	0.7	0	1	1:1	0°	0.73
FC-S-0	0.7	0	1	2:1	0°	0.48
FC-S-0	0.7	0	1	3:1	0°	0.66
FC-S-0	0.7	0	1	4:1	0°	0.62
FC-S-30	0.7	0	1	1:1	0°	0.17
FC-S-30	0.7	0	1	2:1	0°	0.39
FC-S-30	0.7	0	1	3:1	0°	0.48
FC-S-30	0.7	0	1	4:1	0°	0.53
PC-A	0.7	0	1	1:1	0°	0.67
PC-A	0.7	0	1	2:1	0°	0.42

1 inch = 2.54 cm

Table 5. Summary of outlet control entrance-loss coefficients, K_e, for low flows—*Continued*.

Model	Slope (percent)	Fillets (inches)	Number of Barrels	Span:Rise	Skew	Unsubmerged K_e *
PC-A	0.7	0	1	3:1	0°	0.80
PC-A	0.7	0	1	4:1	0°	0.69
FC-S-0	0.7	0	1	3:1	30°	0.84
FC-T-30	0.7	0	3	1:1	0°	0.47
FC-T-0	0.7	0	3	1:1	0°	0.86
FC-T-30	0.7	0	3	1:1	15°	0.43
FC-T-30	0.7	0	3	1:1	30°	0.85
FC-T-30	0.7	0	3	1:1	45°	0.9

* These values are not to be used because of the resolution limits of the pressure transducers.   1 inch = 2.54 cm

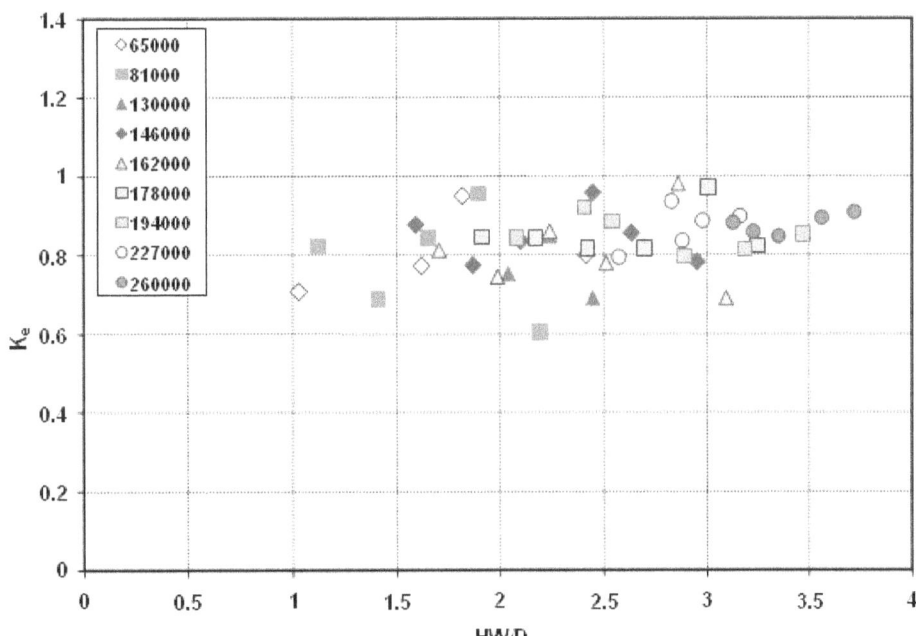

Figure 71. Graph. Entrance loss coefficient versus the Reynolds number, HDS-5 8/3.

The computed entrance loss coefficients scattered between 0.6 and 0.95 for low Reynolds numbers, and the scatter generally decreased with increasing Reynolds numbers, but there was no correlation between K_e and the Reynolds number. Figure 72 shows that the standard deviation of the K_e data gradually decreased with an increasing Reynolds number. This figure is included to demonstrate statistically that, although there is a wide spread in K_e around the 162,000 Reynolds number, the scatter in K_e values decreased as the Reynolds number increased.

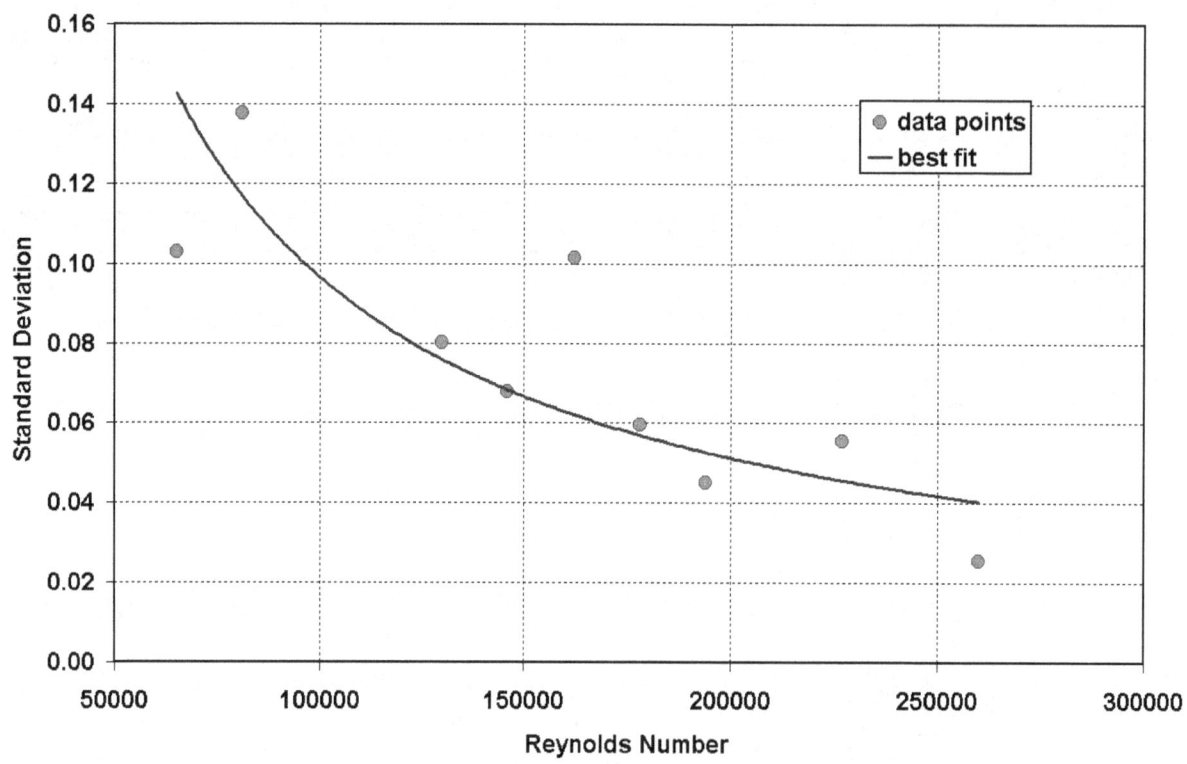

Figure 72. Graph. Standard deviation of K_e versus the Reynolds number.

Based on difficulties measuring low flow depths accurately, the FHWA Hydraulics Laboratory is developing an optical pressure measurement (OPM) system, which is expected to have a much higher resolution (± 0.1 mm) (±0.039 inch). The OPM system will use an array of standpipes mounted along a culvert barrel. Each standpipe has contact image sensors attached, which will measure the water column using imaging techniques. This new sensor was not available for this study but will be used in future research after it has been tested and calibrated.

Another potential concept for determining entrance loss coefficients for low flows is to relate them to a contraction coefficient. The form loss, or head loss H_{Le} or contraction loss H_{Lc}, because of contraction is primarily caused by the reexpansion of the flow following contraction and can therefore be calculated approximately by using areas A and A_c, illustrated in figure 73, and the expansion loss equation (figure 74).

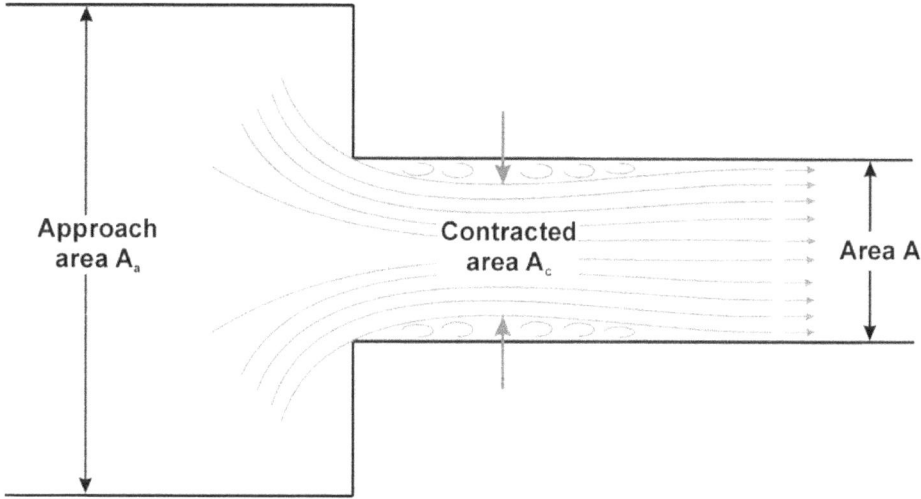

Figure 73. Diagram. Culvert contraction.

By considering the flow expansion from the contracted section, area A_c, to the normal section, area A, the head loss or contraction loss is given by the equation in figure 74.

$$H_{Le} = H_{Lc} = \left(\frac{A}{A_c} - 1\right)^2 \left(\frac{V'^2}{2g}\right) = \left(\frac{A}{C_c A} - 1\right)^2 \left(\frac{V'^2}{2g}\right) = \left(\frac{1}{C_c} - 1\right)^2 \left(\frac{V'^2}{2g}\right)$$

Figure 74. Equation. Expansion loss equation.

The contraction coefficient C_c is equal to A_c / A, and depends on the area ratio A / A_a, the nature of the contraction or inlet geometry, and slightly on the Reynolds number. To quantify the contracted area, detailed velocity profiles can be measured in the contraction zone using either PIV or laser doppler anemometry (LDA). Based on the measured velocity vector fields, streamlines can be integrated to compute the contracted area.

Neither of the latter two concepts has yet been tried because they were beyond the scope of this study, but they have promise.

OUTLET CONTROL EXIT LOSS COEFFICIENTS K_o

The exit loss is a function of the change in velocity at the outlet of the culvert barrel. Figure 75 contains the equation, from HDS-5, for the exit loss for a sudden expansion, as at an end wall.

$$H_{Lo} = 1.0 \left(\frac{V^2}{2g} - \frac{V_d^2}{2g}\right)$$

Figure 75. Equation. Exit loss, with coefficient of 1.

The channel velocity downstream of the culvert is V_d. The mean flow is V. HDS-5 states that the coefficient 1 may overestimate exit loss and a multiplier less then one can be used. If the downstream velocity is neglected, the exit loss is assumed to be the full-flow velocity head in the barrel.

Outlet loss data was analyzed by averaging the pressure tap measurements for the five pressure taps located downstream of the culvert outlet, as illustrated in figures 78, 79, and 80, to determine the average HGL in the tailbox. The downstream velocity was computed from the equation in figure 76. The area of flow was assumed to be the full width of the tailbox multiplied by the average tailwater depth at the five pressure taps.

$$V_d = \frac{Q}{W_{TB} \; TW}$$

Figure 76. Equation. Downstream velocity.

Where:

Q   is discharge.
W_{TB} is the full width of the tailbox.
TW   is the average tailwater depth at the five pressure taps.

The velocity head $V_d^2/2g$ was added to the HGL elevation to establish an average downstream EGL elevation, which was subtracted from the EGL at the end of the culvert to determine the head loss at the outlet, H_{Lo}. The exit loss coefficient, K_o, was then computed from the equation in figure 77.

$$H_{Lo} = K_o \left[\left(\frac{V^2}{2g} \right) - \left(\frac{V_d^2}{2g} \right) \right]$$

Figure 77. Equation. Exit loss, with coefficient K_o.

The difficulty with this procedure is that the downstream velocity, and therefore the computed coefficient, is a function of the laboratory dimensions of the tailbox. The downstream velocity was artificially low because the assumed flow area included a significant dead zone on each side of the culvert. That explains why some of the K_o values in table 6 are actually greater than 1.0.

In an effort to develop a more rational procedure for simulating downstream computation from a laboratory experimental setup, the velocity distributions in the tailbox were measured for a several experiments. The goal was to analyze how velocity profiles expand when moving further downstream and how the downstream velocity affects the exit loss. It would be reasonable to use the effective velocities in the expansion zone, illustrated in figures 78, 79, and 80, at each of the pressure taps to determine EGL. EGL could then be projected to the plane of the culvert outlet to determine the exit head loss, H_{Lo}, as illustrated in figure 80. The effective velocities vary with the distance downstream of the culvert; the challenge is in deciding what velocity to use in the equation in figure 77 to compute K_o. An arbitrary, but reasonable, selection could be the velocity at three culvert widths downstream of the outlet.

Although this alternative procedure was not used in this study to compute the exit loss coefficients, the measurements in the expansion zone provide good insight about flow downstream of culvert outlets.

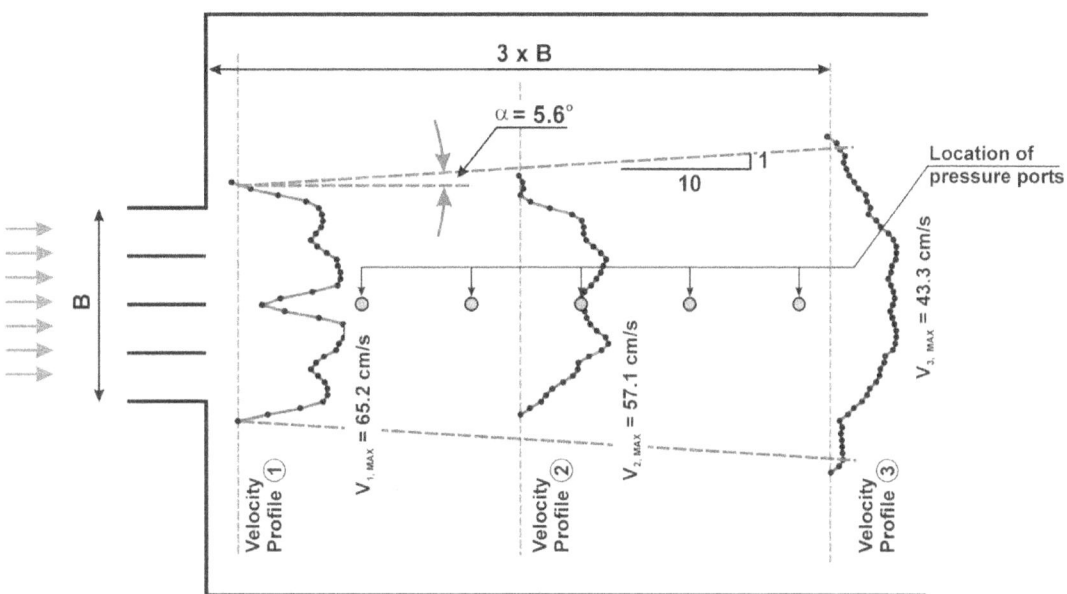

Figure 78. Diagram. Flow expansion in the tailbox for high tailwater.

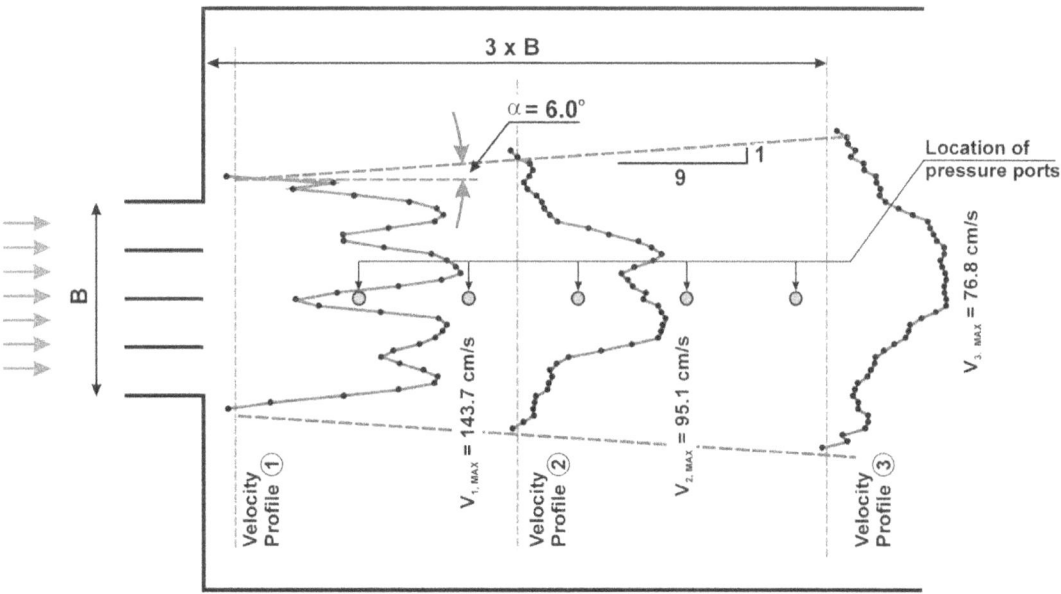

Figure 79. Diagram. Flow expansion in the tailbox for low tailwater.

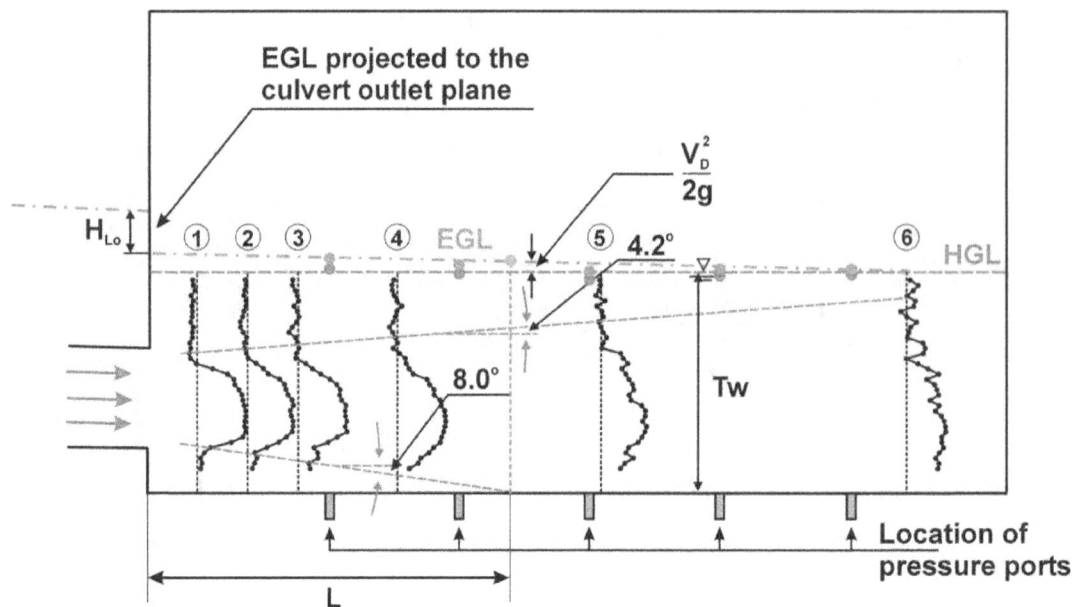

Note: Vertical to horizontal scale is exaggerated by approximately 2:1.

Figure 80. Diagram. Vertical flow expansion in the tailbox and projected EGL.

The energy exit loss coefficients K_o are summarized in table 6 for different culvert barrel configurations. Each K_o value listed in table 6 is an average of several inlet geometries that had the same barrel configuration because, in the opinion of the researchers, the exit loss should not be significantly influenced by the inlet geometry.

Table 6. Summary of outlet loss coefficients.

Culvert barrel configuration		K_o unsubmerged*	K_o submerged
a.		0.73	1.12
b.		1.00	1.11
c.		0.89	0.96
d.		0.94	1.08
e.		1.07	1.19
f.		1.04	1.03
g.		0.66	0.97
h.		1.28	1.35
i.		0.85	1.11
j.	Skew	0.86	0.99
k.	Skew	1.10	1.26

* These values are not to be used because of the resolution limits of the pressure transducers.

FIFTH-ORDER POLYNOMIALS

In the area between submerged and unsubmerged flow conditions, a transition area exists for which neither the submerged nor the unsubmerged forms of the equations provide accurate headwater predictions. An example of this transition area can be seen in figure 81. Fifth-order polynomial equations were developed that predict headwater in this region of uncertainty and over the entire range of measured HW/D ratios.

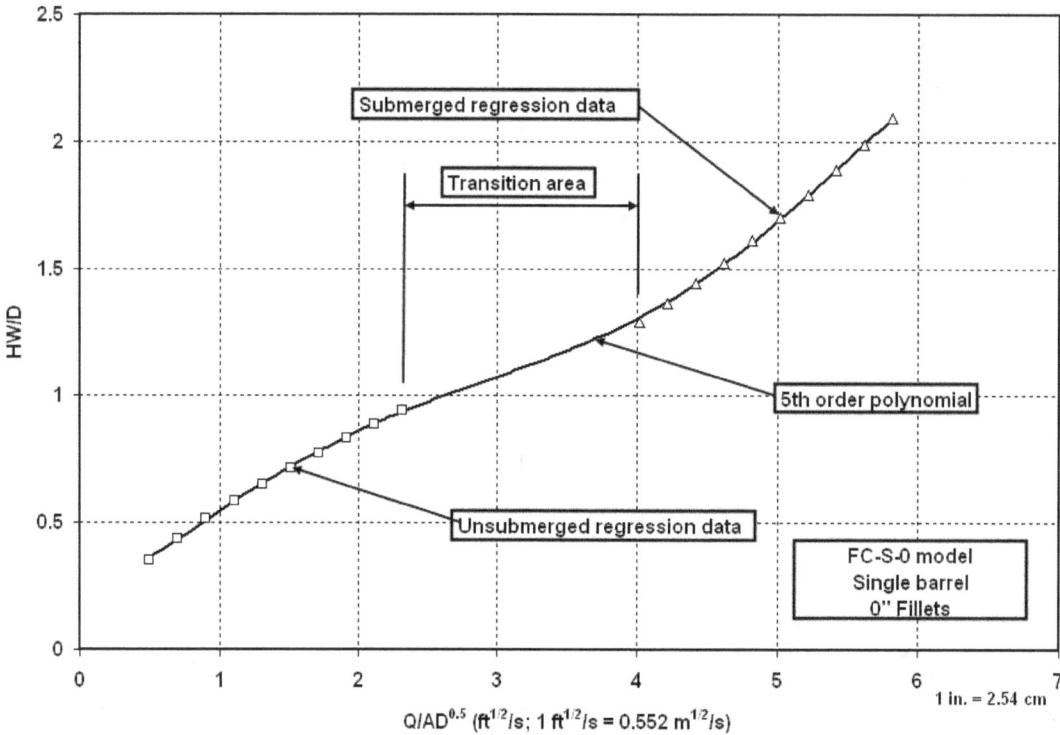

Figure 81. Graph. Transition area, unsubmerged and submerged inlet flow conditions.

For culvert discharges within the range of the regression analysis, the polynomial equation gives a direct solution for inlet headwater, regardless of whether the inlet is submerged:

$$\frac{HW_i}{D} = a + b\left[\frac{Q}{AD^{0.5}}\right] + c\left[\frac{Q}{AD^{0.5}}\right]^2 + d\left[\frac{Q}{AD^{0.5}}\right]^3 + e\left[\frac{Q}{AD^{0.5}}\right]^4 + f\left[\frac{Q}{AD^{0.5}}\right]^5$$

Figure 82. Equation. Transition area, unsubmerged and submerged inlet flow conditions.

Where:

a to f   are polynomial regression design coefficients.

The polynomial regression coefficients are presented in tables 7 to 10. Application of the polynomial equations presents numerical difficulties and errors at low and high values of HW/D.

At low values, the HW/D prediction approaches the intercept coefficient, a, instead of zero, as it should. At high values of HW/D, greater than about 2.3, the equations have a maximum. Therefore, the useful operating range of the polynomial equations is approximately $0.4 < HW/D < 2.3$.

Table 7. Polynomial regression coefficients, models tested for bevels and corner fillets.

Model	Fillet (inches)	Span: Rise	a	b	c	d	e	f
FC-S-0	0	1:1	0.211536	0.224341	0.208370	−0.121994	0.024676	−0.001609
FC-S-0	6	1:1	0.224471	0.247312	0.186514	−0.113509	0.023189	−0.001514
FC-S-0	12	1:1	0.245464	0.218175	0.210202	−0.121260	0.024063	−0.001536
PC-A	0	1:1	0.194217	0.310678	0.109365	−0.077409	0.016183	−0.001059
PC-A	6	1:1	0.203074	0.313529	0.107677	−0.076856	0.016046	−0.001046
PC-A	12	1:1	0.210576	0.314554	0.101951	−0.072650	0.014939	−0.000956
PC-A	6	2:1	0.186778	0.482282	−0.070441	−0.003094	0.003186	−0.000258
PC-A	12	2:1	0.189232	0.496842	−0.090840	0.005876	0.001525	−0.000152

1 inch = 2.54 cm

Table 8. Polynomial regression coefficients, models tested for span-to-rise ratio.

Model	Fillet (inches)	Span: Rise	a	b	c	d	e	f
FC-S-0	0	1:1	0.211536	0.224341	0.208370	−0.121994	0.024676	−0.001609
FC-S-0	0	2:1	0.207152	0.432143	−0.011800	−0.032349	0.009224	−0.000668
FC-S-0	0	3:1	0.227185	0.361402	0.077047	−0.069984	0.015731	−0.001063
FC-S-0	0	4:1	0.246621	0.315820	0.126140	−0.091438	0.019626	−0.001311
FC-S-30	0	1:1	0.163450	0.127103	0.256193	−0.131631	0.025211	−0.001601
FC-S-30	0	2:1	0.114704	0.376884	−0.007409	−0.024267	0.006977	−0.000502
FC-S-30	0	3:1	0.141479	0.321334	0.064086	−0.056126	0.012732	−0.000862
FC-S-30	0	4:1	0.230000	0.117000	0.241000	−0.126000	0.025000	−0.001640
PC-A	0	1:1	0.194217	0.310678	0.109365	−0.077409	0.016183	−0.001059
PC-A	0	2:1	0.154724	0.592825	−0.190065	0.048610	−0.006357	0.000370
PC-A	0	3:1	0.200747	0.424467	0.000339	−0.032906	0.008306	−0.000569
PC-A	0	4:1	0.220017	0.404031	0.029449	−0.046577	0.010869	−0.000734

1 inch = 2.54 cm

Table 9. Polynomial regression coefficients, models tested for multiple barrels.

Model	Fillet (inches)	No. of Barrels	a	b	c	d	e	f
FC-S-0	6	1	0.224471	0.247312	0.186514	−0.113509	0.023189	−0.001514
FC-S-30	6	1	0.148953	0.250437	0.129334	−0.080435	0.016465	−0.001069
FC-D-0	6	2	0.154694	0.436581	−0.034324	−0.017500	0.006033	−0.000449
FC-D-0-E	6	2	0.168455	0.406534	−0.000921	−0.031807	0.008541	−0.000602
FC-D-30	6	2	0.101044	0.419241	−0.055078	−0.002691	0.002885	−0.000237
FC-D-30-E	6	2	0.103678	0.402060	−0.035969	−0.011102	0.004465	−0.000338
FC-T-0	6	3	0.185570	0.425924	−0.015139	−0.027376	0.007816	−0.000559
FC-T-0-E	6	3	0.220522	0.370427	0.057110	−0.060032	0.013803	−0.000936
FC-T-30	6	3	0.132482	0.340923	0.042681	−0.044257	0.010089	−0.000674
FC-T-30-E	6	3	0.146450	0.369886	0.018327	−0.037160	0.009256	−0.000639
FC-Q-0	6	4	0.154223	0.448350	−0.040698	−0.014061	0.005238	−0.000392
FC-Q-0-E	6	4	0.202572	0.369396	0.056302	−0.059162	0.013704	−0.000934
FC-Q-30	6	4	0.108394	0.355482	0.027300	−0.034330	0.007929	−0.000526
FC-Q-30-E	6	4	0.144926	0.350135	0.041407	−0.046111	0.010763	−0.000730
PC-A	12	1	0.210576	0.314554	0.101951	−0.072650	0.014939	−0.000956
PC-B	12	2	0.099284	0.607914	−0.221590	0.066777	−0.009847	0.000567
PC-B-E	12	2	0.149381	0.540449	−0.142259	0.029009	−0.002577	0.000099
PC-C	12	3	0.147326	0.545578	−0.146179	0.032208	−0.003485	0.000151
PC-C-E	12	3	0.153250	0.555777	−0.152070	0.035094	−0.004187	0.000219
PC-D	12	4	0.107009	0.600450	−0.207455	0.060926	−0.008924	0.000516
PC-D-E	12	4	0.158898	0.585920	−0.191810	0.048483	−0.005859	0.000294

1 inch = 2.54 cm

Table 10. Polynomial regression coefficients, models tested for skewed headwall.

Model	Fillet (inches)	Skew Angle (degrees)	a	b	c	d	e	f
FC-S-30	0	30	0.213123	0.685460	−0.298162	0.088583	−0.012236	0.000663
FC-T-30	0	0	0.128409	0.355047	0.026694	−0.037498	0.008897	−0.000603
FC-T-0	0	0	0.183389	0.409080	0.005613	−0.035825	0.009284	−0.000650
FC-T-30	0	15	0.182031	0.686256	−0.277043	0.082113	−0.011673	0.000654
FC-T-30	0	30	0.225978	0.658427	−0.240324	0.063449	−0.007969	0.000409
FC-T-30	0	45	0.187459	0.743672	−0.323642	0.103601	−0.016120	0.000958

1 inch = 2.54 cm

74

CHAPTER 7. CONCLUSIONS AND RECOMMENDATIONS

Since most highway designers use HY-8 or a similar program for hydraulic design of culverts, highway agencies do not receive much benefit from research results until the results are finally coded into a computer program. Even the Corps of Engineers Hydrologic Engineering Center River Analysis System (HECRAS) program, used for most water surface profile studies, includes the HY-8 logic for culvert evaluations. The HY-8 program is currently being revised to incorporate recent research results and to allow for user-defined design coefficients.

As it is currently coded, HY-8 does not allow for user-defined design coefficients, which restricts practitioners from using research results until they are finally coded into a program. Since HY-8 and other similar programs are based on HDS-5, the best current approach for advancing the research to an implementation stage is to develop coefficients and performance curves in a format that is equivalent to the HDS-5 format, to derive fifth-order polynomials to facilitate coding inlet control performance curves in a computer program, and to identify which results will have the most impact on the applicability or clarity of HDS-5.

HDS-5 is the most widely recognized publication available on culvert hydraulics. All of the inlets tested in this study were inlets covered in HDS-5 with various combinations of entrance improvements. HDS-5 does not include thumbnail sketches of the inlet configurations that go with the descriptions, and it is sometimes difficult to visualize some of the details of the inlets. Nevertheless, experimental results in this study were compared to published coefficients in HDS-5 in cases where the inlets appeared to match the descriptions in HDS-5.

FINDINGS AND CONCLUSIONS

The major findings and conclusions of this study are:

- The discharge intensity is the primary independent variable used in culvert hydraulic analyses. As it is defined in HDS-5, the discharge intensity unnecessarily has units of $ft^{1/2}/s$, but it could just as easily be defined as a dimensionless Froude number by including the acceleration of gravity in the denominator. Almost all other parameters in culvert hydraulics are dimensionless and to make discharge intensity also dimensionless would greatly facilitate converting from one system to another in this period of dual units.

- The 20.32-cm- (8-inch-) radius top plate bevel was the optimum shape among six shapes tested. That radius is the full wall thickness of the top plate. The optimum top plate bevel does improve culvert performance significantly. The improvement is more pronounced for multiple barrels at higher headwater depths.

- The 45-degree straight top plate bevel, used for SDDOT field cast inlets, is an improvement over the square-edge top plate specified in HDS-5 for concrete box culverts with 0-degree wingwall flare angles.

- The precast models with 0-degree wingwall flare and optimum curved top plate bevels, as tested in these experiments, performed consistently better than the comparable field cast models with 0-degree wingwall flare and the traditional 45-degree top plate bevel. The precast models performed between the 0- and 30-degree-flared wingwall field cast models, as illustrated in figures 83 and 84, except for multiple barrels at HW/D ratio greater than 1.5. At those ratios, the precast models actually performed better than the 30-degree-flared wingwall field cast model.

- The rounded bevels for wingwall top edges had no discernible effect on culvert performance. The square-edge models performed as well as the models with rounded bevels.

- The size of corner fillets had no discernible effect on the performance curves provided the net culvert area was used in the computation of the discharge intensity. The performance curves associated with various corner fillets in figures 46 and 47 can reasonably be combined as single curves, as illustrated in figure 85. The inlet control tests showed no difference for the various corner fillet sizes. There was a slight difference in the outlet control coefficient for the 30.48-cm (12-inch) corner fillets, but that difference can probably be attributed to experimental scatter.

- Multiple barrels had very little effect on performance curves for the field cast models. The double, triple, and quad curves in figures 50 and 51 can reasonably be combined as single curves, as illustrated in figures 86 and 87, and they could further be combined with the single barrel curves without much loss in accuracy. This observation gives credibility to the common practice of using single barrel design coefficients for multiple barrel culverts.

- Multiple barrels had more pronounced effect on performance curves for the precast models with the optimum top bevels, especially when the HW/D ratio was greater than 1.5, as illustrated in figure 88. This was an unexpected result, and several tests were rerun to confirm it. Most highway agencies, however, design culverts for headwater depth ratios below 1.5. Consequently, the practice of using single barrel coefficients is still reasonable.

- Span-to-rise ratios greater than 1:1 had very little effect on performance curves for either the field cast or the precast models, as illustrated in figures 89 to 91. Nevertheless, the error in using the 1:1 design coefficients for wider span culverts tends to be on the unconservative side, especially for inlets with flared wingwalls, as illustrated in figure 90.

76

- Extending the inner walls onto the approach apron for multiple barrel culverts had no discernible effect on performance curves. There is no hydraulic advantage (or disadvantage) from extending the inner walls.

- Skewed headwalls are unavoidable for some highway alignments, but they do have a detrimental effect on performance curves, as illustrated in figure 92. Results from this study do not agree with the limited guidance in HDS-5 for skewed headwalls.

- Exit flow from culvert models expanded at a gradual expansion angle of 5 to 6 degrees for a significant distance downstream from the culvert for both low and high tailwater depths.

- Entrance loss coefficients for low flows are important for fish passage considerations, but the unacceptable scatter, which can be attributed to instrumentation limitations for the very small losses that were associated with low flows, makes those results unreliable. Consequently, those coefficients for outlet control with free surface flow are not among the recommended results from this study. They are tabulated in an appendix for the benefit of other researchers who may have similar difficulties.

- A commonly used prefabricated inlet for small circular culverts was highlighted from the literature review. Although it was not tested in this study, results were fairly consistent in two separate studies from the literature, and design coefficients from those studies are considered reliable.

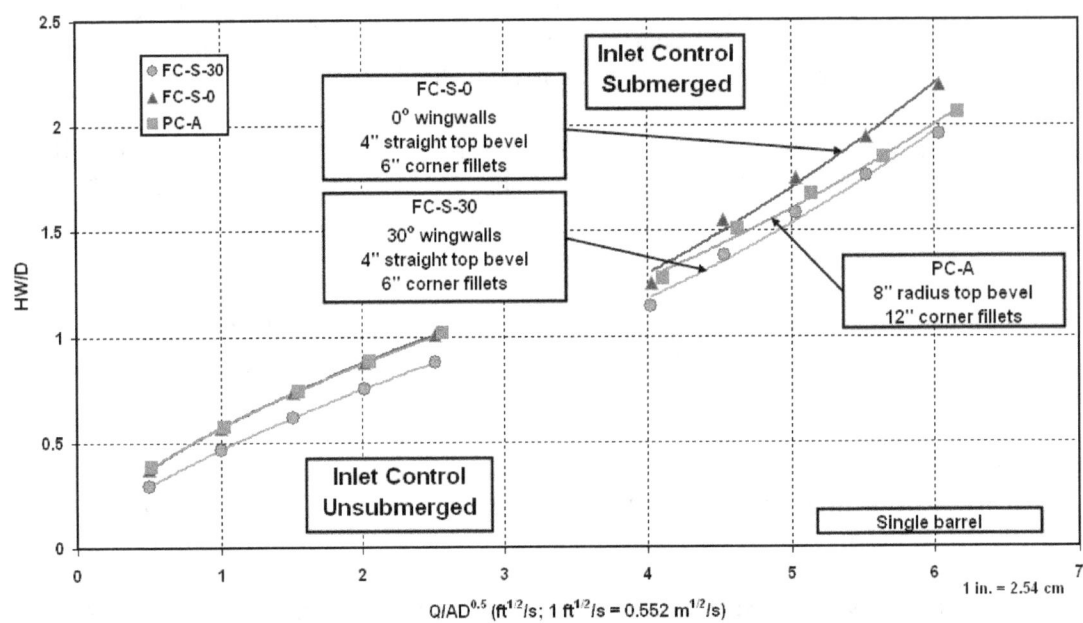

**Figure 83. Graph. PC and FC single barrel models
(sketches 1, 7, 11 in figure 93).**

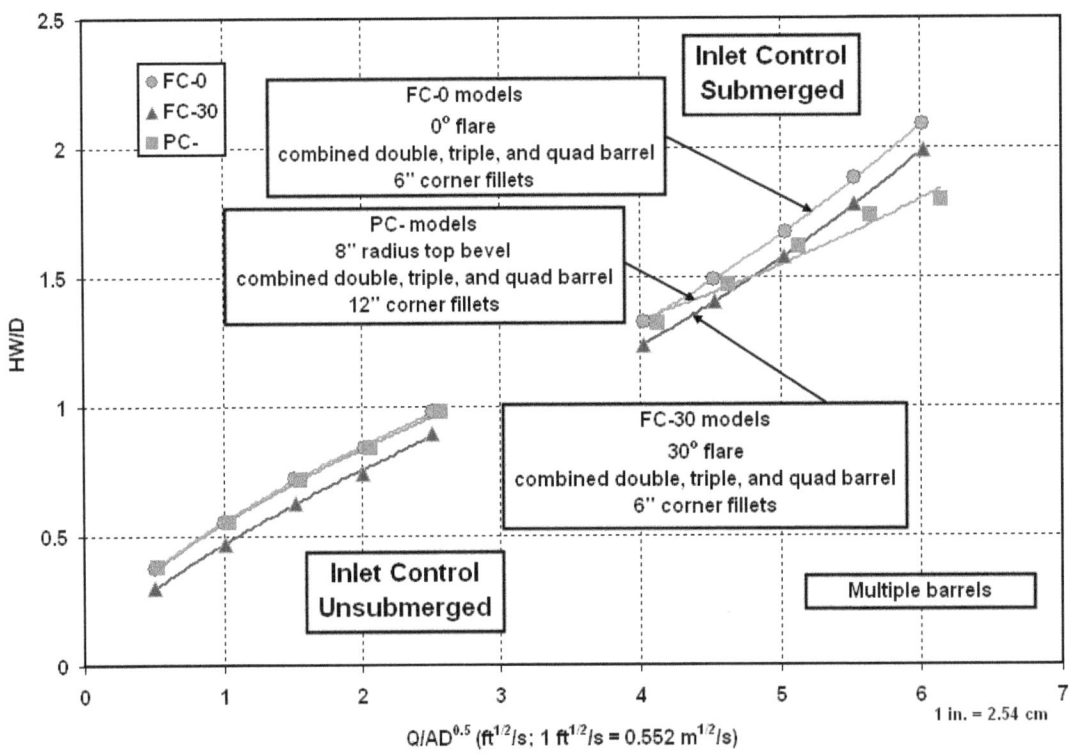

**Figure 84. Graph. PC and FC multiple barrel models
(sketches 1, 2, 7, 8, 11, 12 in figure 93).**

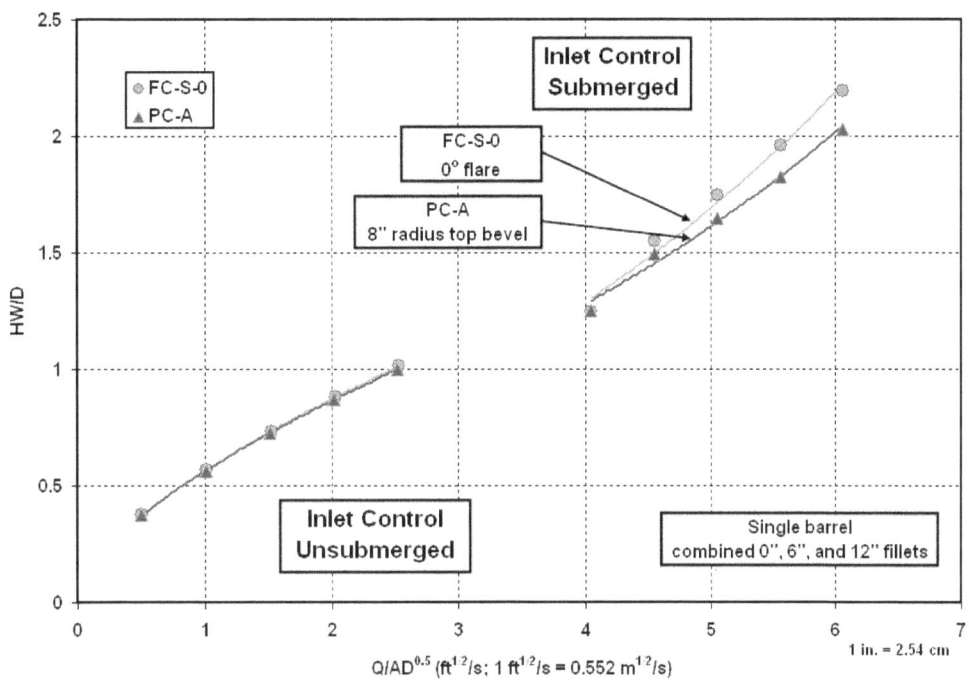

Figure 85. Graph. Combined corner fillet data, FC-S-0 and PC-A models (sketches 7, 10, 11, 14 in figure 93).

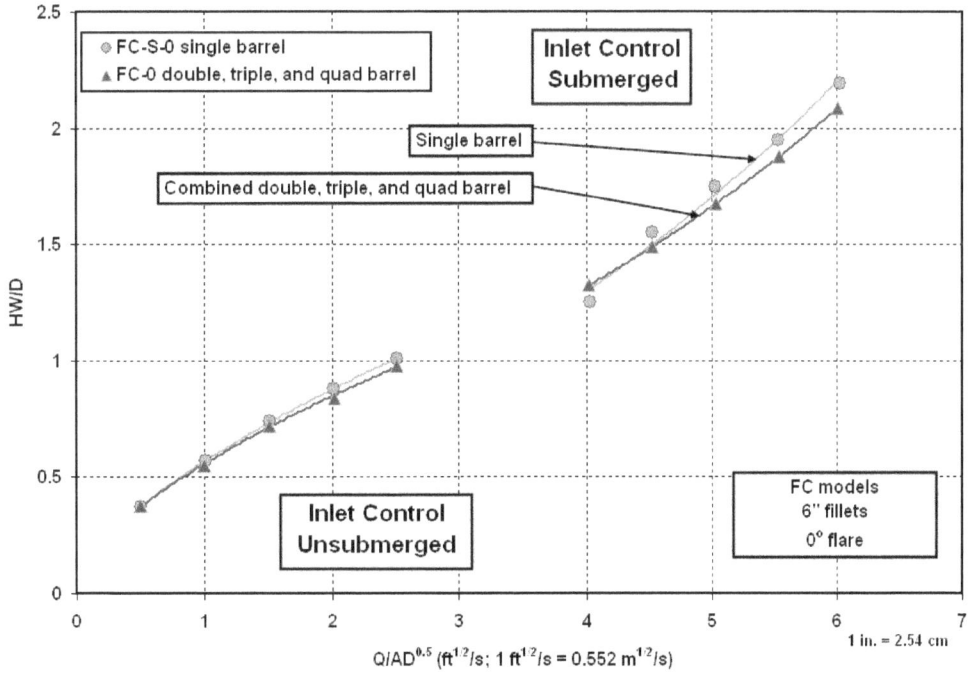

Figure 86. Graph. Combined multiple barrel data, FC-0 models (sketches 7, 8 in figure 93).

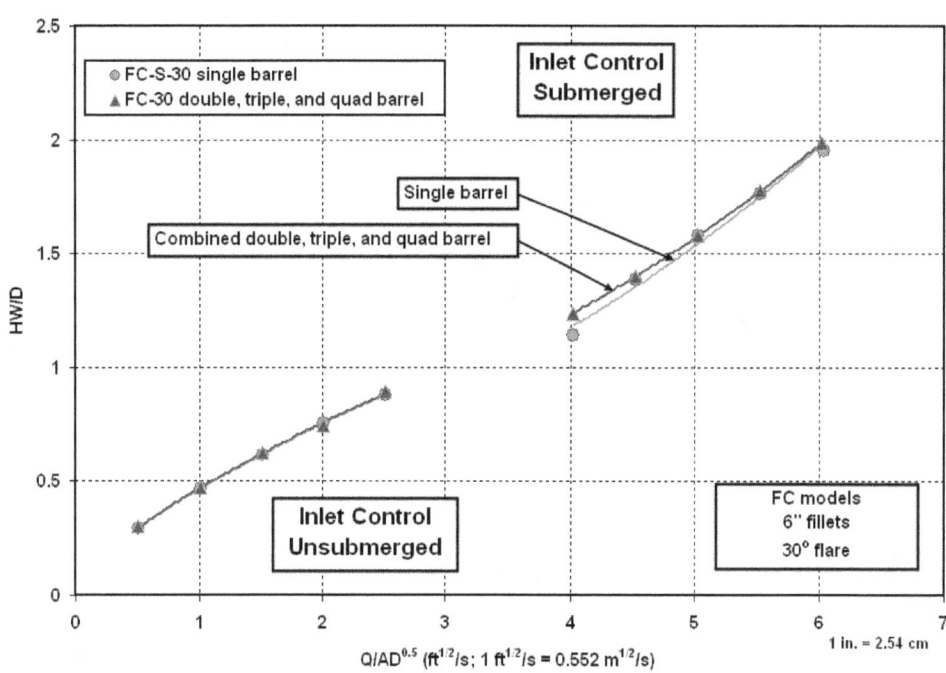

Figure 87. Graph. Combined multiple barrel data, FC-30 models (sketches 1, 2 in figure 93).

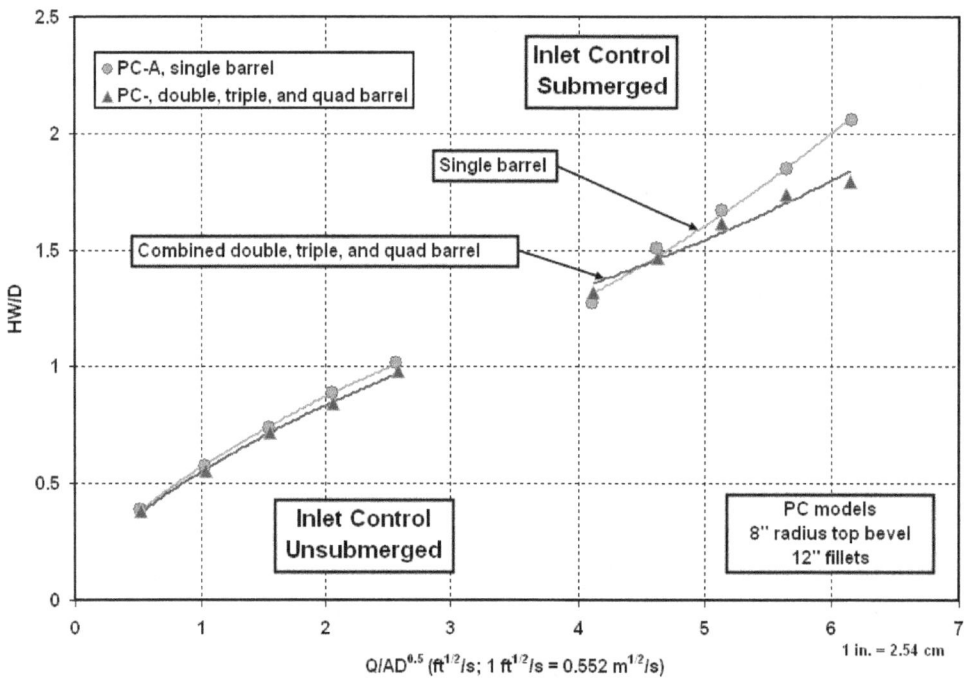

Figure 88. Graph. Combined multiple barrel data, PC models (sketches 11, 12 in figure 93).

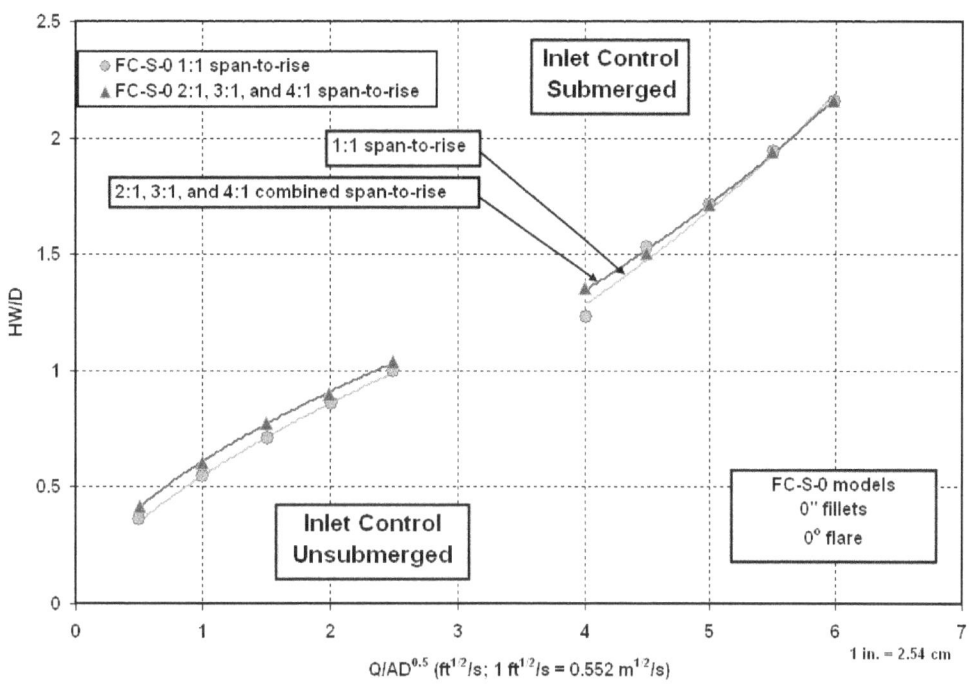

Figure 89. Graph. Combined span-to-rise data, FC-S-0 models (sketches 7, 9 in figure 93).

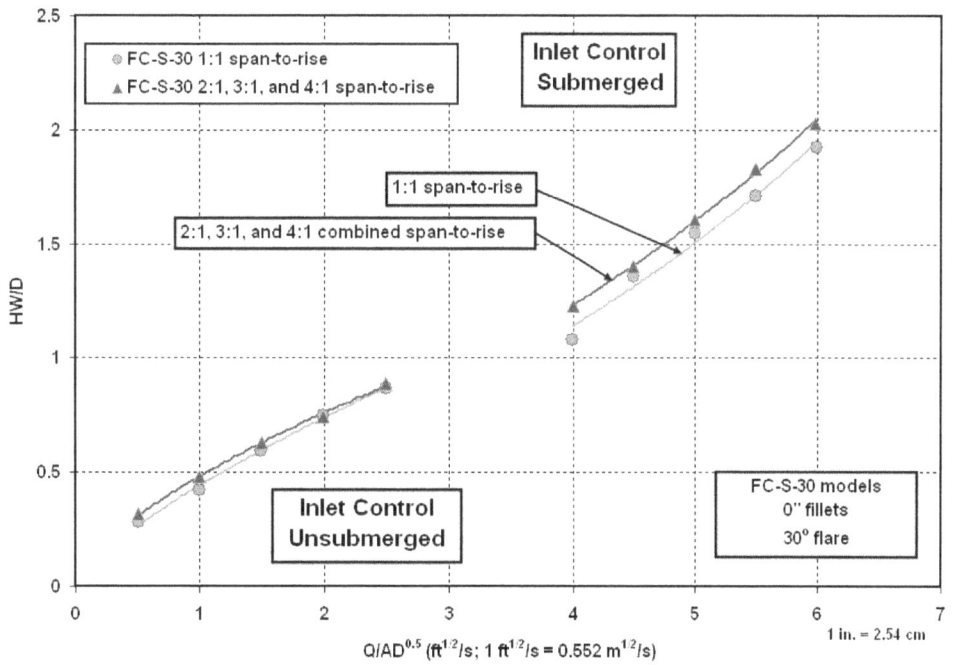

Figure 90. Graph. Combined span-to-rise data, FC-S-30 models (sketches 1, 3 in figure 93).

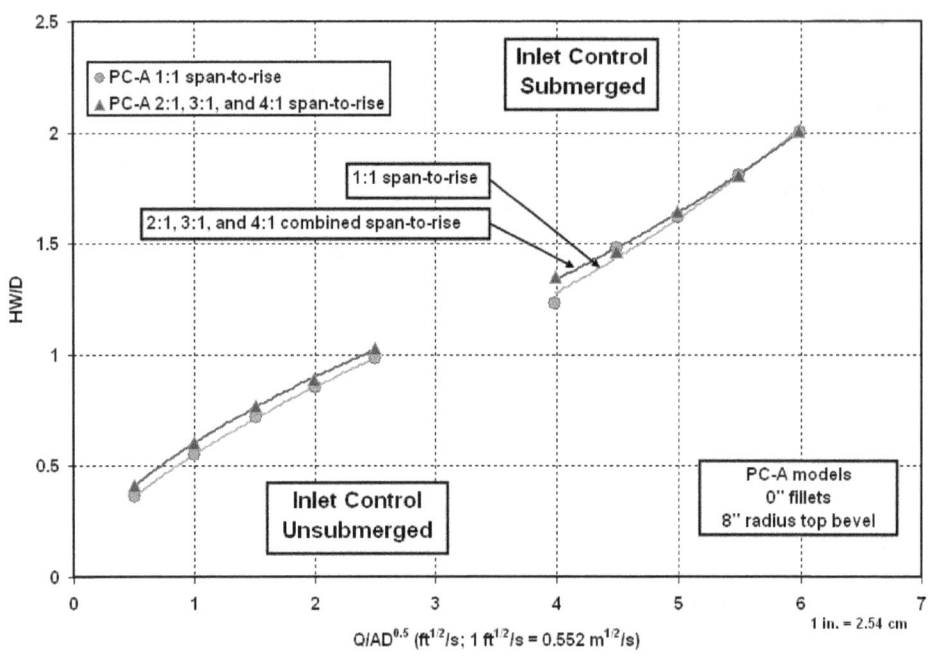

Figure 91. Graph. Combined span-to-rise data, PC models (sketches 10, 13 in figure 93).

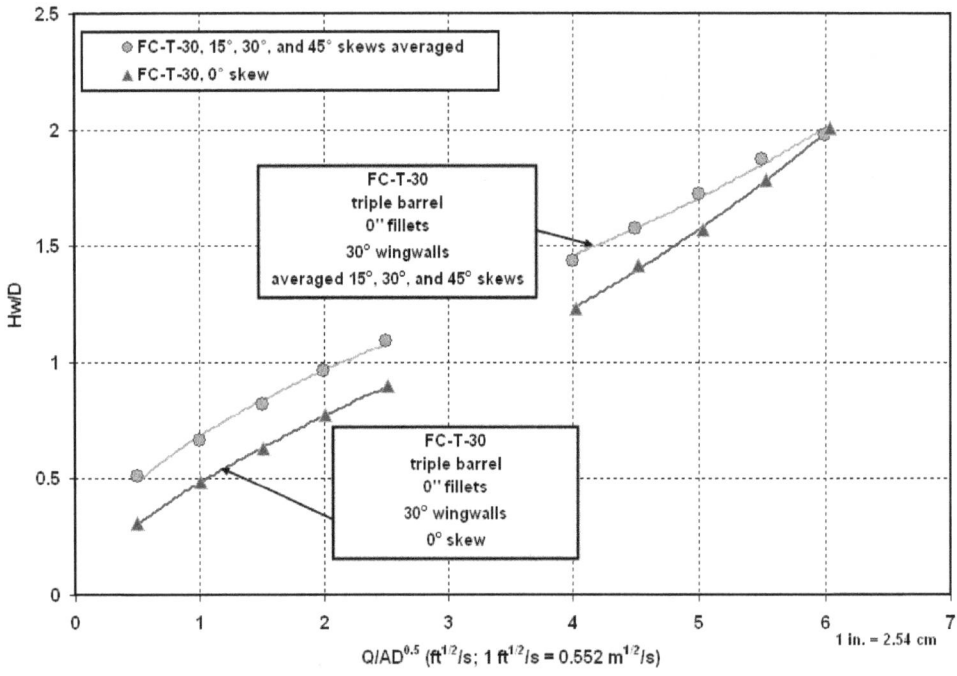

Figure 92. Graph. Skewed and nonskewed headwalls, FC-T-30 models (sketches 4, 5 in figure 93).

82

RECOMMENDATIONS

The following recommendations are for consideration by practitioners involved in field cast and precast culvert installations:

- Use the 20.32-cm- (8-inch-) radius top plate bevel for precast inlets where it is feasible to use more detailed forms.

- Continue the practice of using straight chamfer on the top edge of wingwalls because testing showed no benefit from rounding these edges.

- Base decisions to extend inner walls of multiple barrel culverts onto the apron on issues—such as debris control or aesthetics—other than hydraulics because testing showed no hydraulic advantage or disadvantage to extending these walls.

The following recommendations are for consideration as enhancements to future editions of hydraulic design manuals such as HDS-5 and future generations of hydraulic design software such as HY-8:

- Redefine discharge intensity as a dimensionless parameter, $Q/A(gD)^{0.5}$, to facilitate conversion from one system of units to another. Keep the acceleration of gravity, g, in the independent parameter rather than embedding it in the design coefficients.

- Use thumbnail sketches, similar to those used in this report (figure 93), to clarify which inlet configurations correspond to design coefficients. Thumbnail sketches should also indicate the test conditions embankment slope if the applicable range of design coefficients might be affected.

- Define culvert area as net area rather than gross area when corner fillets are used. This should not affect current coefficients in HDS-5 because most historical testing was done without considering corner fillets.

- Expand the table of design coefficients to include box culverts with 30-degree-flared wingwalls and 45-degree straight beveled top plates, as described for sketch 1 in table 11. This expansion would reflect the performance of SD's field cast culverts with flared wingwalls.

- Expand the table of design coefficients to include box culverts with 0-degree-flared wingwalls and 45-degree straight beveled and rounded top plates, as described for sketches 6 and 10 in table 11. This expansion would reflect the performance of SD field cast and precast culverts with straight-mitered inlets.

- Expand the table of design coefficients to include multiple barrel box culverts, as described for sketches 2, 8, and 12 in table 11. This expansion would document that multiple barrel culverts were tested and had very little effect on performance.

- Expand the table of design coefficients to include wide-span box culverts, as described for sketches 3, 9, and 13 in table 11. This expansion would document that, for flared wingwall installations, wide-span box culverts are slightly less effective than culverts with 1:1 span-to-rise ratios.

- Use the design coefficients in table 11 for inlets skewed to the flow direction due to skewed highway alignment relative to the flow direction. More research is needed to develop design coefficients for culvert barrels that are skewed to the flow direction. Although skewing culvert barrels to the flow direction is considered "bad practice," research is needed to show just how bad it is because the practice is not uncommon.

- Use table 12 for fifth-order polynomial coefficients that correspond to design coefficients in table 11. Fifth-order polynomials are coding expedients for computer software.

Table 11. Design coefficients suggested for future editions of HDS-5.

Description	Source	K_e	K_1	M_1	K_2	M_2	c	Y
Box, reinforced concrete								
30° to 75° flared wingwalls	Chart 8/1, HDS-5				0.469	0.696	0.033	0.751
30° to 75° flared wingwalls, Square-edged crown	Table 12, HDS-5	0.4						
30° to 75° flared wingwalls, Crown edge rounded or top edge beveled	Table 12, HDS-5	0.2						
30° flared wingwalls, top edge beveled at 45°								
Single barrel	Sketch 1, figure 93	0.26	0.005	1.05	0.44	0.74	0.04	0.48
2, 3, and 4 multiple barrels	Sketch 2, figure 93	0.32			0.47	0.68	0.04	0.62
2:1 to 4:1 span-to-rise ratio	Sketch 3, figure 93	0.20			0.48	0.65	0.041	0.57
15° skewed headwall, multiple barrels	Sketch 4, figure 93	0.36			0.69	0.49	0.029	0.95
30° to 45° skewed headwall, multiple barrels	Sketch 5, figure 93	0.45			0.69	0.49	0.027	1.02
0° flared wingwalls, extended sides								
Square-edged at crown	Chart 8/3, HDS-5	0.7			0.55	0.64	0.05	0.55
Square-edged at crown	Sketch 6, figure 93	0.79	0.055	0.68	0.55	0.64	0.047	0.55
45° straight bevel at crown, 0- and 6-inch corner fillets	Sketch 7, figure 93	0.48			0.56	0.62	0.045	0.55
45° straight bevel at crown, 2, 3, and 4 multiple barrels	Sketch 8, figure 93	0.52			0.55	0.59	0.038	0.69
45° straight bevel at crown, 2:1 to 4:1 span-to-rise ratio	Sketch 9, figure 93	0.37			0.61	0.57	0.041	0.67
Crown rounded at 8-inch radius, 0- and 6-inch corner fillets	Sketch 10, figure 93	0.24			0.56	0.62	0.038	0.67
Crown rounded at 8-inch radius, 12-inch corner fillets	Sketch 11, figure 93	0.3			0.56	0.62	0.038	0.67
Crown rounded at 8-inch radius, 12-inch corner fillets, 2, 3, and 4 multiple barrels	Sketch 12, figure 93	0.54			0.55	0.60	0.023	0.96
Crown rounded at 8-inch radius, no fillets, 2:1 to 4:1 span-to-rise ratio	Sketch 13, figure 93	0.30			0.61	0.57	0.033	0.79

Note: The source column refers to two documents: HDS-51 and figure 93 in this report. Empty cells mean no data are available.   1 inch = 2.54 cm

85

Table 12. Fifth-order polynomial coefficients.

Description	Source	a	b	c	d	e	f
Box, reinforced concrete							
30° to 75° flared wingwalls	Chart 8/1, HDS–5	0.175897	0.20233	0.139818	-0.07197	0.014148	-0.00093
30° to 75° flared wingwalls, Square-edged crown	Table 12, HDS–5						
30° to 75° flared wingwalls, Crown edge rounded or top edge beveled	Table 12, HDS–5						
30° flared wingwalls, top edge beveled at 45°							
Single barrel	Sketch 1, figure 93	0.163450	0.127103	0.256193	-0.131630	0.025211	-0.001600
2, 3, and 4 multiple barrels	Sketch 2, figure 93	0.112542	0.375074	0.002657	-0.026380	0.006867	-0.000470
2:1 to 4:1 span-to-rise ratio	Sketch 3, figure 93	0.182681	0.209471	0.158774	-0.092970	0.019381	-0.001310
15° skewed headwall, multiple barrels	Sketch 4, figure 93	0.182031	0.686256	-0.277040	0.0821130	-0.011670	0.000654
30° to 45° skewed headwall, multiple barrels	Sketch 5, figure 93	0.363958	0.283523	0.069040	-0.044221	0.008965	-0.000595
0° flared wingwalls, extended sides							
Square-edged at crown	Chart 8/3, HDS–5						
Square-edged at crown	Sketch 6, figure 93	0.278122	-0.002930	0.448521	-0.228310	0.044973	-0.00299
45° straight bevel at crown, 0- and 6-inch corner fillets	Sketch 7, figure 93	0.244295	0.129322	0.339620	-0.189660	0.038635	-0.0026
45° straight bevel at crown, 2, 3, and 4 multiple barrels	Sketch 8, figure 93	0.164261	0.43842	-0.03128	-0.01919	0.006288	-0.00046
45° straight bevel at crown, 2:1 to 4:1 span-to-rise ratio	Sketch 9, figure 93	0.254968	0.273934	0.154712	-0.09911	0.020506	-0.00135
Crown rounded at 8-inch radius, 0- and 6-inch corner fillets	Sketch 10, figure 93	0.203424	0.31092	0.10783	-0.07609	0.015779	-0.00102
Crown rounded at 8-inch radius, 12-inch corner fillets	Sketch 11, figure 93	0.203424	0.31092	0.10783	-0.07609	0.015779	-0.00102
Crown rounded at 8-inch radius, 12-inch corner fillets, 2, 3, and 4 multiple barrels	Sketch 12, figure 93	0.10315	0.619895	-0.23147	0.071093	-0.01075	0.000628
Crown rounded at 8-inch radius, no fillets, 2:1 to 4:1 span-to-rise ratio	Sketch 13, figure 93	0.199715	0.446748	-0.02414	-0.02334	0.006761	-0.00048

Notes: The source column refers to two documents: HDS–51 and figure 93 in this report. Empty cells mean no data are available.

1 inch = 2.54 cm

a. Sketch 1 30°-flared wingwalls; top edge beveled at 45°	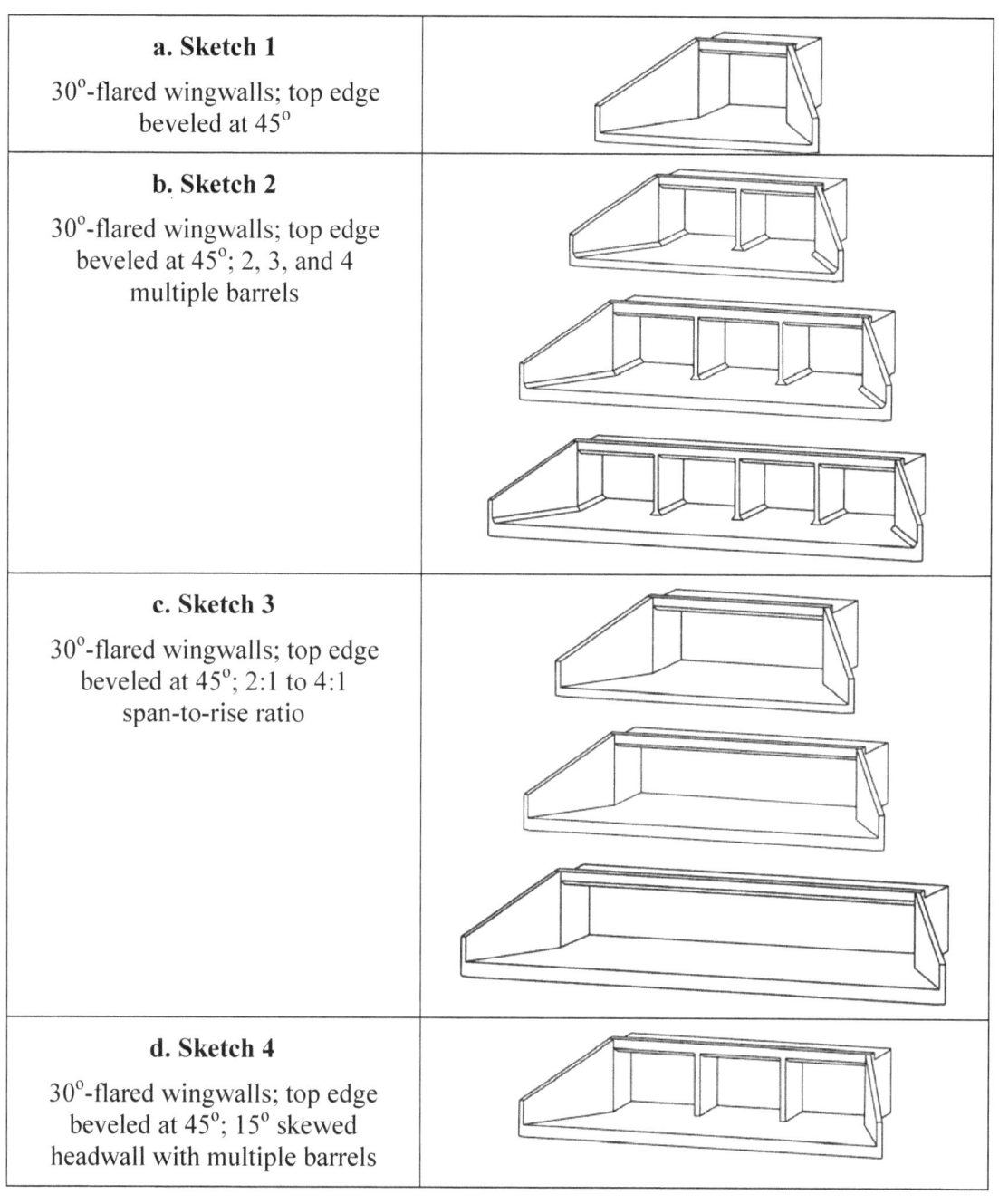
b. Sketch 2 30°-flared wingwalls; top edge beveled at 45°; 2, 3, and 4 multiple barrels	
c. Sketch 3 30°-flared wingwalls; top edge beveled at 45°; 2:1 to 4:1 span-to-rise ratio	
d. Sketch 4 30°-flared wingwalls; top edge beveled at 45°; 15° skewed headwall with multiple barrels	

Figure 93. Thumbnail sketches of inlets recommended for implementation.

e. Sketch 5 30°-flared wingwalls; top edge beveled at 45°; 30° to 45° skewed headwall with multiple barrels	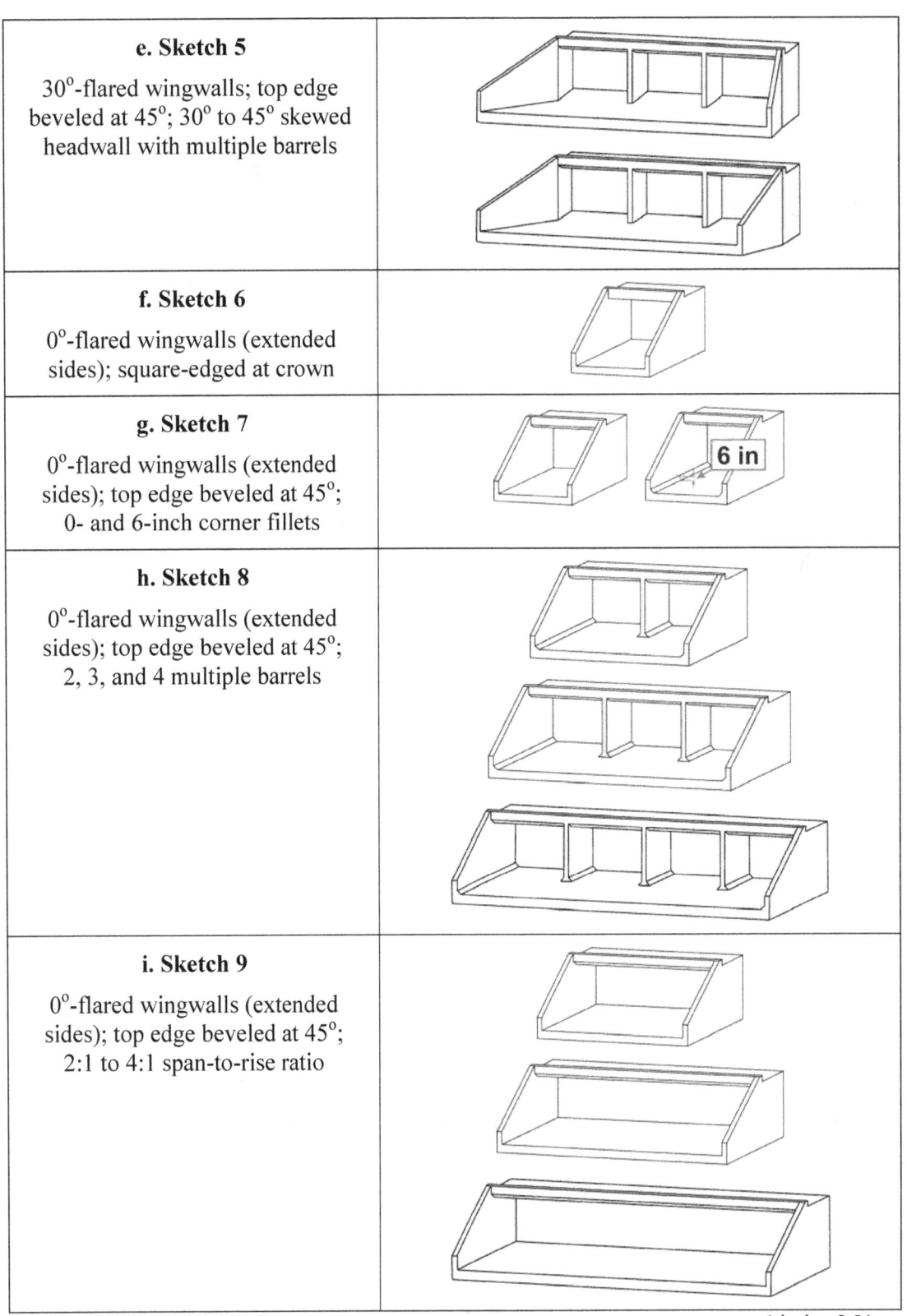
f. Sketch 6 0°-flared wingwalls (extended sides); square-edged at crown	
g. Sketch 7 0°-flared wingwalls (extended sides); top edge beveled at 45°; 0- and 6-inch corner fillets	
h. Sketch 8 0°-flared wingwalls (extended sides); top edge beveled at 45°; 2, 3, and 4 multiple barrels	
i. Sketch 9 0°-flared wingwalls (extended sides); top edge beveled at 45°; 2:1 to 4:1 span-to-rise ratio	

1 inch = 2.54 cm

Figure 93. Thumbnail sketches of inlets recommended for implementation—*Continued.*

88

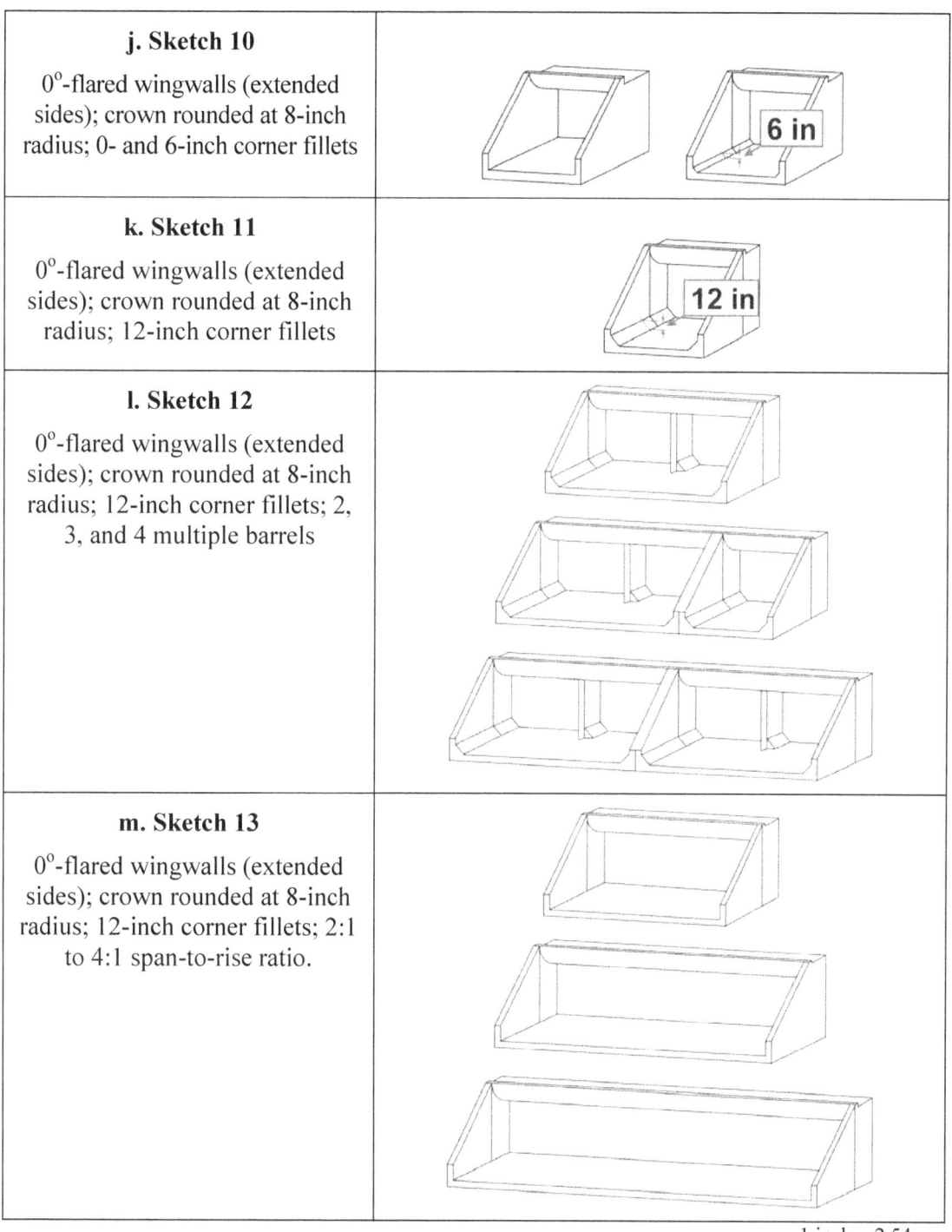

j. Sketch 10 0°-flared wingwalls (extended sides); crown rounded at 8-inch radius; 0- and 6-inch corner fillets	
k. Sketch 11 0°-flared wingwalls (extended sides); crown rounded at 8-inch radius; 12-inch corner fillets	
l. Sketch 12 0°-flared wingwalls (extended sides); crown rounded at 8-inch radius; 12-inch corner fillets; 2, 3, and 4 multiple barrels	
m. Sketch 13 0°-flared wingwalls (extended sides); crown rounded at 8-inch radius; 12-inch corner fillets; 2:1 to 4:1 span-to-rise ratio.	

1 inch = 2.54 cm

Figure 93. Thumbnail sketches of inlets recommended for implementation—*Continued.*

n. Sketch 14 0°-flared wingwalls (extended sides); top edge beveled at 45°; 12-inch corner fillets	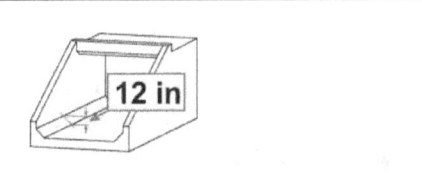

Note: Extended center walls do not affect flow conditions.   1 inch = 2.54 cm

Figure 93. Thumbnail sketches of inlets recommended for implementation—*Continued.*

APPENDIX A. EXPANDED TEST MATRIX

This appendix is an expanded description of the test matrix. The expanded matrix includes sketches, more detailed information about test conditions, and extra tests that were conducted in response to questions posed during the study. The matrix was posted in the lab and used as a checklist during the experimental phase.

Figure 30 in chapter 5 contains the seven different bevel edge configurations that were tested in the miniflume to determine the best edge configuration. The criterion to determine the best edge configuration was the contracted distance outside the viscous boundary layer (effective flow depth at the vena contracta, also illustrated in figure 30). PIV data were used to analyze the flow at the culvert entrance and to quantify the effective flow depth.

Tables 13 through 16 summarize tests performed in the culvert test facility to analyze the effects of bevels, multiple barrels, span-to-rise ratios, and skewed headwalls. Two extra inlet geometries—an FC-S-0 hybrid and a PC-A hybrid—were tested in response to questions on the draft final report regarding the wingwall bevels and are included in table 16. The FC-S-0 hybrid was a combination of the field cast top bevel with the precast wingwall bevels, and the PC-A hybrid was a combination of the precast top bevel with the field cast square edge wingwalls.

Table 13. Tests to analyze the effects of bevels for wingwalls and top edges.

Inlet	Model ID	Sketch	Wingwall flare angle	Bevels	Corner fillets (inches)	Barrel size (feet)	Culvert slopes	Tailwater
1.1	FC-S-0		0°	4-inch-straight top bevel, no WW bevel	0	6 x 6	0.03, 0.007	High, Low
1.2	FC-S-0		0°	4-inch-straight top bevel, no WW bevel	6	6 x 6	0.03, 0.007	High, Low
1.3	FC-S-0		0°	4-inch-straight top bevel, no WW bevel	12	6 x 6	0.03, 0.007	High, Low
1.4	PC-A		0°	8-inch-radius top bevel, 4-inch-radius WW bevels	0	6 x 6	0.03, 0.007	High, Low
1.5	PC-A		0°	8-inch-radius top bevel, 4-inch-radius WW bevels	6	6 x 6	0.03, 0.007	High, Low
1.6	PC-A		0°	8-inch-radius top bevel, 4-inch-radius WW bevels	12	6 x 6	0.03, 0.007	High, Low
1.7	PC-A		0°	8-inch-radius top bevel, 4-inch-radius WW bevels	6	6 x 12	0.03	High, Low
1.8	PC-A		0°	8-inch-radius top bevel, 4-inch-radius WW bevels	12	6 x 12	0.03	High, Low
1.9	FC-S-0 Hybrid		0°	4-inch-straight top bevel, 4-inch-radius WW bevels	0	6 x 6	0.03	High, Low

1 inch = 2.54 cm; 1 ft = 0.305 m

Table 13. Tests to analyze the effects of bevels for wingwalls and top edges—*Continued*.

Inlet	Model ID	Sketch	Wingwall flare angle	Bevels	Corner fillets (inches)	Barrel size (feet)	Culvert slopes	Tailwater
1.10	PC-A Hybrid		0°	8-inch-radius top bevel, no WW bevels	0	6 x 6	0.03	High, Low

1 inch = 2.54 cm; 1 ft = 0.305 m

Notes: Target discharge intensities for unsubmerged flow, $Q/AD^{0.5}$ = 0.5, 1.0, 2.0, 3.5, and 4.0 ($ft^{0.5}$/s). Target discharge intensities for submerged flow, $Q/AD^{0.5}$ = 4.5, 5.0, 5.5, and 6.0 ($ft^{0.5}$/s). One $ft^{0.5}$/s equals 0.552 $m^{0.5}$/s. Contraction ratio of headbox width to total span of culvert model was held constant at 2.67. WW is wingwalls.

Table 14. Tests to analyze the effects of multiple barrels.

Inlet	Model ID	Sketch	WW flare angle	Bevels	Barrel size (feet)	Culvert slopes	Corner fillets (inches)	Barrels	Inner wall
2.1	FC-S-0		0°	4-inch-straight top bevel, no WW bevel	6 x 6	0.03, 0.007	6	1	None
2.2	FC-S-30		30°	4-inch-straight top bevel, no WW bevel	6 x 6	0.03, 0.007	6	1	None
2.3	FC-D-0		0°	4-inch-straight top bevel, no WW bevel	6 x 6	0.03, 0.007	6	2	Not extended
2.4	FC-D-0-E		0°	4-inch-straight top bevel, no WW bevel	6 x 6	0.03, 0.007	6	2	Extended
2.5	FC-D-30		30°	4-inch-straight top bevel, no WW bevel	6 x 6	0.03, 0.007	6	2	Not extended
2.6	FC-D-30-E		30°	4-inch-straight top bevel, no WW bevel	6 x 6	0.03, 0.007	6	2	Extended
2.7	FC-T-0		0°	4-inch-straight top bevel, no WW bevel	6 x 6	0.03, 0.007	6	3	Not extended
2.8	FC-T-0-E		0°	4-inch-straight top bevel, no WW bevel	6 x 6	0.03, 0.007	6	3	Extended
2.9	FC-T-30		30°	4-inch-straight top bevel, no WW bevel	6 x 6	0.03, 0.007	6	3	Not extended
2.10	FC-T-30-E		30°	4-inch-straight top bevel, no WW bevel	6 x 6	0.03, 0.007	6	3	Extended
2.11	FC-Q-0		0°	4-inch-straight top bevel, no WW bevel	6 x 6	0.03, 0.007	6	4	Not extended

1 inch = 2.54 cm; 1 ft = 0.305 m

Table 14. Tests to analyze the effects of multiple barrels—*Continued.*

Inlet	Model ID	Sketch	WW flare angle	Bevels	Barrel size (feet)	Culvert slopes	Corner fillets (inches)	Barrels	Inner wall
2.12	FC-Q-0-E		0°	4-inch-straight top bevel, no WW bevel	6 x 6	0.03, 0.007	6	4	Extended
2.13	FC-Q-30		30°	4-inch-straight top bevel, no WW bevel	6 x 6	0.03, 0.007	6	4	Not extended
2.14	FC-Q-30-E		30°	4-inch-straight top bevel, no WW bevel	6 x 6	0.03, 0.007	6	4	Extended
2.15	PC-A		0°	8-inch-radius top bevel, 4-inch-radius WW bevels	6 x 6	0.03, 0.007	12	1	None
2.16	PC-B		0°	8-inch-radius top bevel, 4-inch-radius WW bevels	6 x 6	0.03, 0.007	12	2	Not extended
2.17	PC-B-E		0°	8-inch-radius top bevel, 4-in.-radius WW bevels	6 x 6	0.03, 0.007	12	2	Extended
2.18	PC-C		0°	8-inch-radius top bevel, 4-inch-radius WW bevels	6 x 6	0.03, 0.007	12	3	Not extended
2.19	PC-C-E		0°	8-inch-radius top bevel, 4-inch-radius WW bevels	6 x 6	0.03, 0.007	12	3	Extended
2.20	PC-D		0°	8-inch-radius top bevel, 4-inch-radius WW bevels	6 x 6	0.03, 0.007	12	4	Not extended
2.21	PC-D-E		0°	8-inch-radius top bevel, 4-inch-radius WW bevels	6 x 6	0.03, 0.007	12	4	Extended

1 inch = 2.54 cm; 1 ft = 0.305 m

Notes: Target discharge intensities for unsubmerged flow, $Q/AD^{0.5}$ = 0.5, 1.0, 2.0, 3.5, and 4.0 ($ft^{0.5}$/s). Target discharge intensities for submerged flow, $Q/AD^{0.5}$ = 4.5, 5.0, 5.5, and 6.0 ($ft^{0.5}$/s). One $ft^{0.5}$/s equals 0.552 $m^{0.5}$/s. Contraction ratio of headbox width to total span of culvert model was held constant at 2.67. WW is wingwalls.

Table 15. Tests to analyze the effects of the span-to-rise ratio.

Inlet	Model ID	Sketch	WW flare angle	Bevels	Barrel size (feet)	Culvert Slopes	Corner fillets (inches)	Span-to-rise
3.1	FC-S-0		0°	4-inch-straight top bevel, no WW bevel	6 x 6	0.03, 0.007	0	1:1
3.2	FC-S-30		30°	4-inch-straight top bevel, no WW bevel	6 x 6	0.03, 0.007	0	1:1
3.3	PC-A		0°	8-inch-radius top bevel, 4-inch-radius WW bevels	6 x 6	0.03, 0.007	0	1:1
3.4	FC-S-0		0°	4-inch-straight top bevel, no WW bevel	6 x 12	0.03, 0.007	0	2:1
3.5	FC-S-30		30°	4-inch-straight top bevel, no WW bevel	6 x 12	0.03, 0.007	0	2:1
3.6	PC-A		0°	8-inch-radius top bevel, 4-inch-radius WW bevels	6 x 12	0.03, 0.007	0	2:1
3.7	FC-S-0		0°	4-inch-straight top bevel, no WW bevel	6 x 18	0.03, 0.007	0	3:1
3.8	FC-S-30		30°	4-inch-straight top bevel, no WW bevel	6 x 18	0.03, 0.007	0	3:1
3.9	PC-A		0°	8-inch-radius top bevel, 4-inch-radius WW bevels	6 x 18	0.03, 0.007	0	3:1
3.10	FC-S-0		0°	4-inch-straight top bevel, no WW bevel	6 x 24	0.03, 0.007	0	4:1

1 inch = 2.54 cm; 1 ft = 0.305 m

Table 15. Tests to analyze the effects of the span-to-rise ratio—*Continued*.

Inlet	Model ID	Sketch	WW flare angle	Bevels	Barrel size (feet)	Culvert slopes	Corner fillets (inches)	Span-to-rise
3.11	FC-S-30		30°	4-inch-straight top bevel, no WW bevel	6 x 24	0.03, 0.007	0	4:1
3.12	PC-A		0°	8-inch-radius top bevel, 4-inch-radius WW bevels	6 x 24	0.03, 0.007	0	4:1

1 inch = 2.54 cm; 1 ft = 0.305 m

Notes: Target discharge intensities for unsubmerged flow, $Q/AD^{0.5}$ = 0.5, 1.0, 2.0, 3.5, and 4.0 ($ft^{0.5}$/s). Target discharge intensities for submerged flow, $Q/AD^{0.5}$ = 4.5, 5.0, 5.5, and 6.0 ($ft^{0.5}$/s). One $ft^{0.5}$/s equals 0.552 $m^{0.5}$/s. Contraction ratio of headbox width to total span of culvert model was held constant at 2.67. WW is wingwalls.

Table 16. Tests to analyze the effects of skew.

Inlet	Model ID	Sketch	WW flare angle	Bevels	Barrel size (feet)	Culvert slopes	Corner fillets (inches)	Barrels	Span-to-rise	Skew
4.1	FC-T-0		0°	4-inch-straight top bevel, no WW bevel	6 x 6	0.03, 0.007	0	3	-	0°
4.2	FC-T-30		30°	4-inch-straight top bevel, no WW bevel	6 x 6	0.03, 0.007	0	3	-	0°
4.3	FC-T-30		30°	4-inch-straight top bevel, no WW bevel	6 x 6	0.03, 0.007	0	3	-	15°
4.4	FC-T-30		30°	4-inch-straight top bevel, no WW bevel	6 x 6	0.03, 0.007	0	3	-	30°
4.5	FC-T-30		30°	4-inch-straight top bevel, no WW bevel	6 x 6	0.03, 0.007	0	3	-	45°
4.6	FC-S-30		30°	4-inch-straight top bevel, no WW bevel	6 x 18	0.03, 0.007	0	-	3:1	30°

1 inch = 2.54 cm; 1 ft = 0.305 m

Notes: Target discharge intensities for unsubmerged flow, $Q/AD^{0.5}$ = 0.5, 1.0, 2.0, 3.5, and 4.0 ($ft^{0.5}$/s). Target discharge intensities for submerged flow, $Q/AD^{0.5}$ = 4.5, 5.0, 5.5, and 6.0 ($ft^{0.5}$/s). One $ft^{0.5}$/s equals 0.552 $m^{0.5}$/s. Contraction ratio of headbox width to total span of culvert model was held constant at 2.67. WW is wingwalls.

APPENDIX B. INLET CONTROL COMPARISON CHARTS

This appendix includes all of the comparison charts used to evaluate the performance curves for various inlets tested under inlet control conditions. Charts from this appendix were selectively pulled into the text of the main body of this report to explain the analysis of data, but some of the charts included in this appendix showed such subtle differences that the authors chose not to overcomplicate the body of the report. Figures 94 through 100 are comparison graphs from experiments on the effects of fillets and bevels. Figures 101 through 122 are comparison graphs from experiments on the effects of multiple barrels. Figures 123 through 129 are comparison graphs from experiments on the effects of the span-to-rise ratio. Figures 130 and 131 are comparison graphs from experiments on the effects of skewed headwalls.

The following charts, figures 94 through 100, show the effects of fillets and bevel.

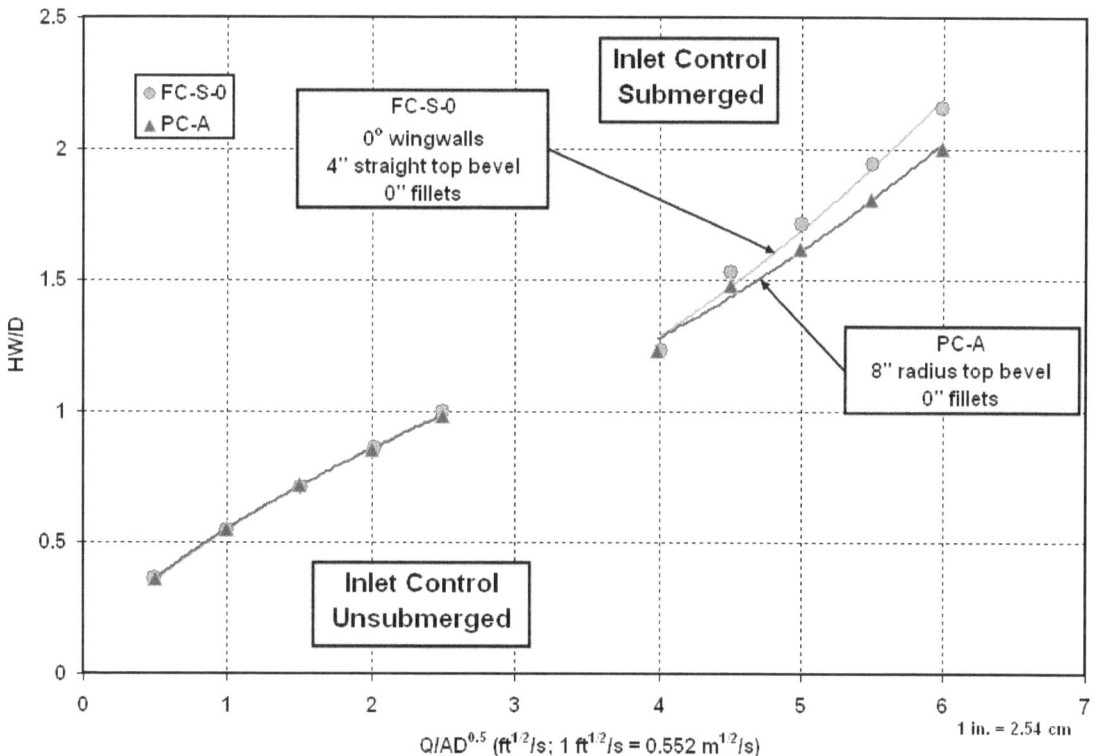

Figure 94. Graph. Inlet control, FC-S-0 and PC-A, no corner fillets.

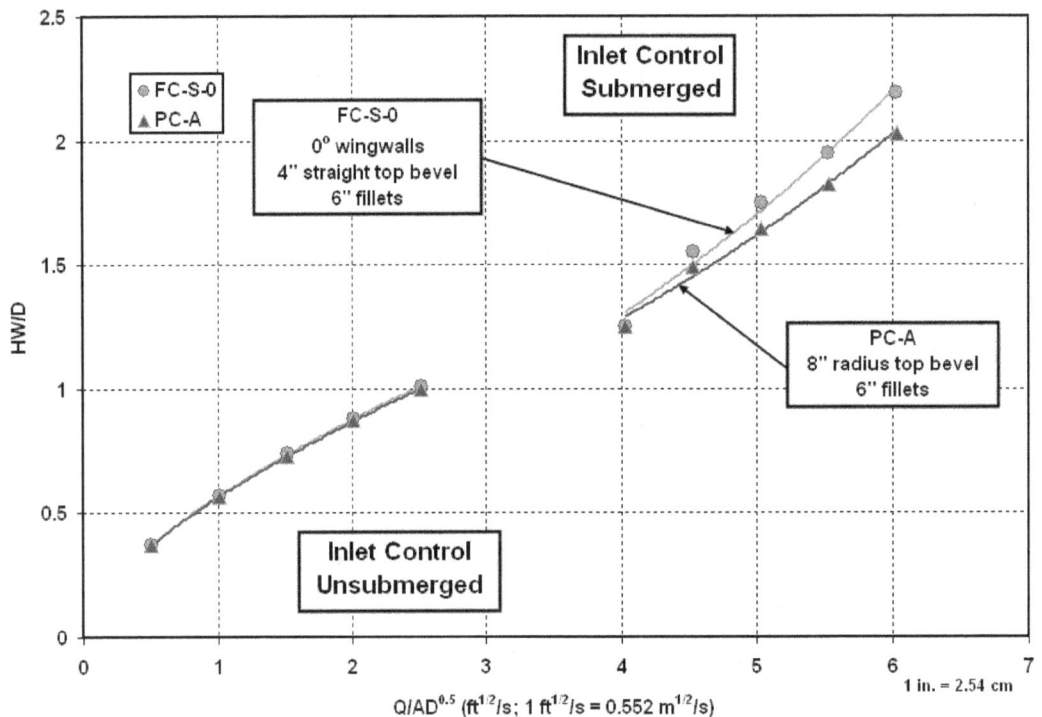

Figure 95. Graph. Inlet control, FC-S-0 and PC-A, 15.24-cm (6-inch) corner fillets.

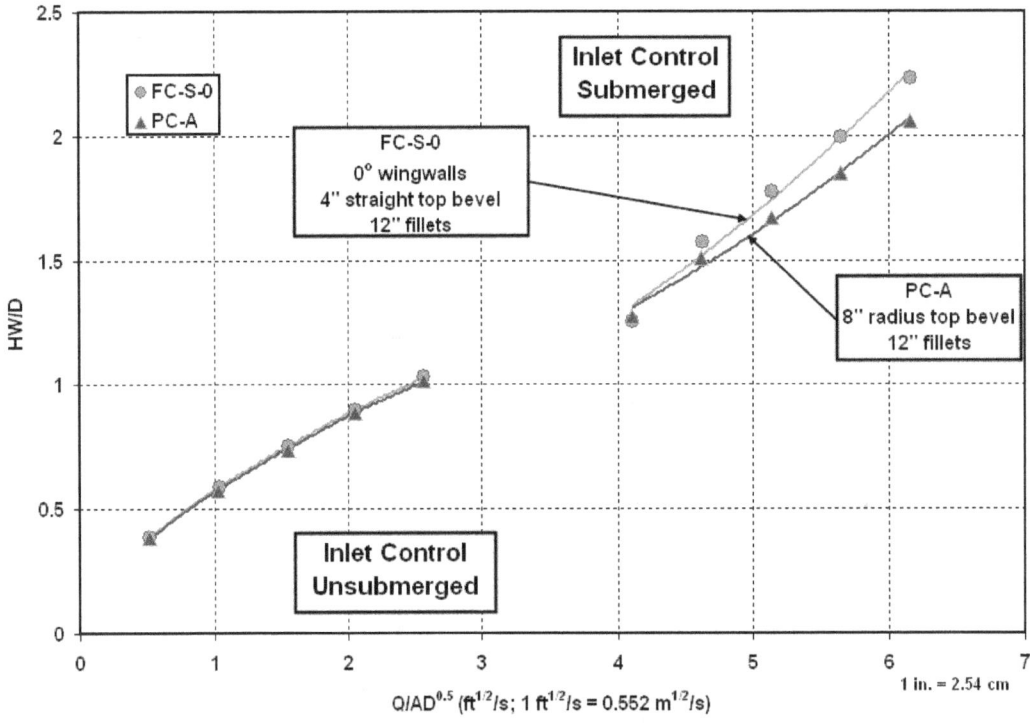

Figure 96. Graph. Inlet control, FC-S-0 and PC-A, 30.48-cm (12-inch) corner fillets.

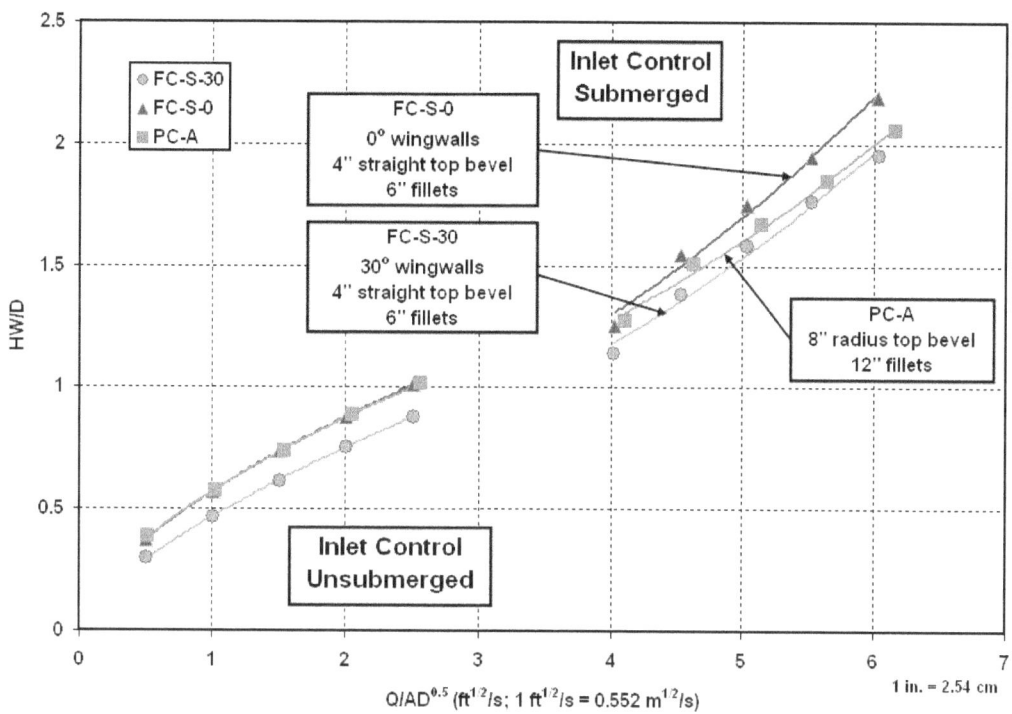

Figure 97. Graph. Inlet control, FC-S-30, FC-S-0, and PC-A.

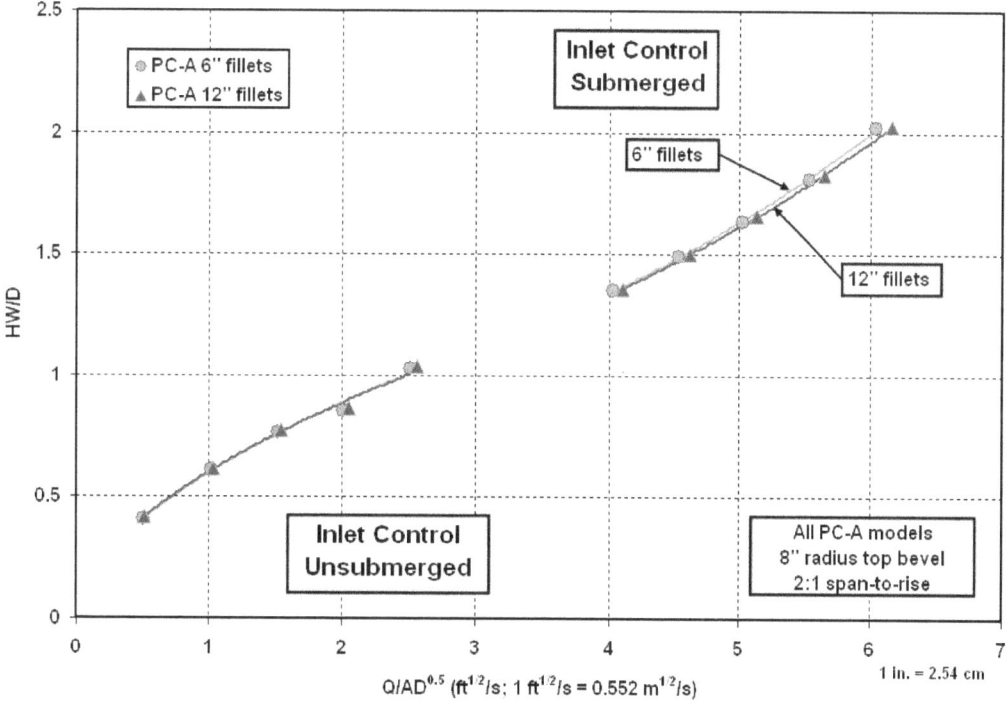

Figure 98. Graph. Inlet control, PC-A, 15.24- and 30.48-cm (6- and 12-inch) corner fillets.

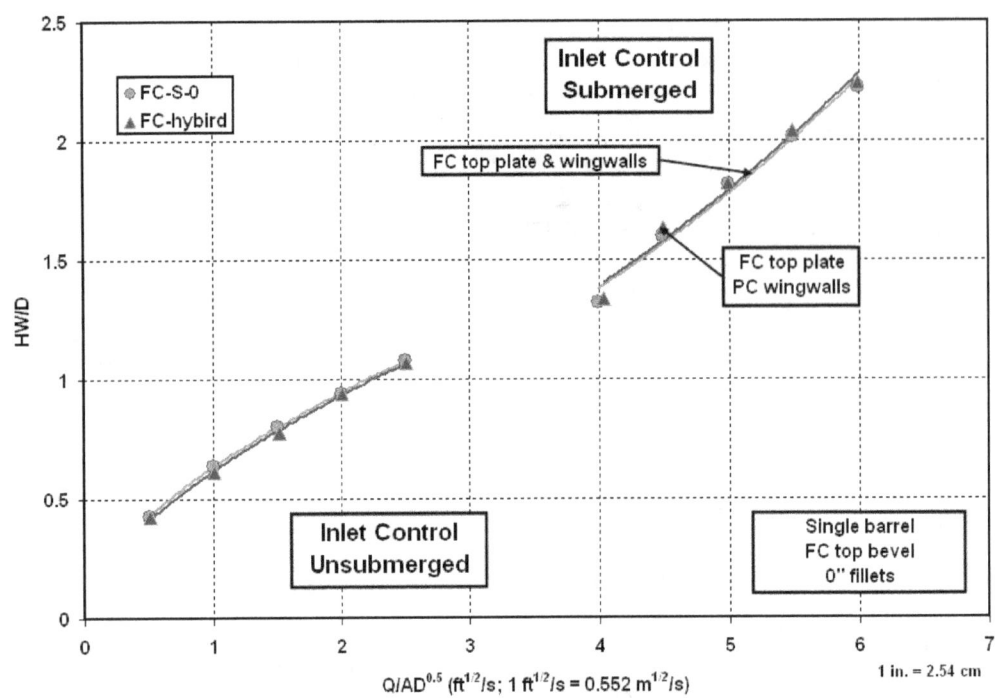

Figure 99. Graph. Inlet control, field cast hybrid inlet with 10.16-cm- (4-inch-) radius bevel on wingwalls.

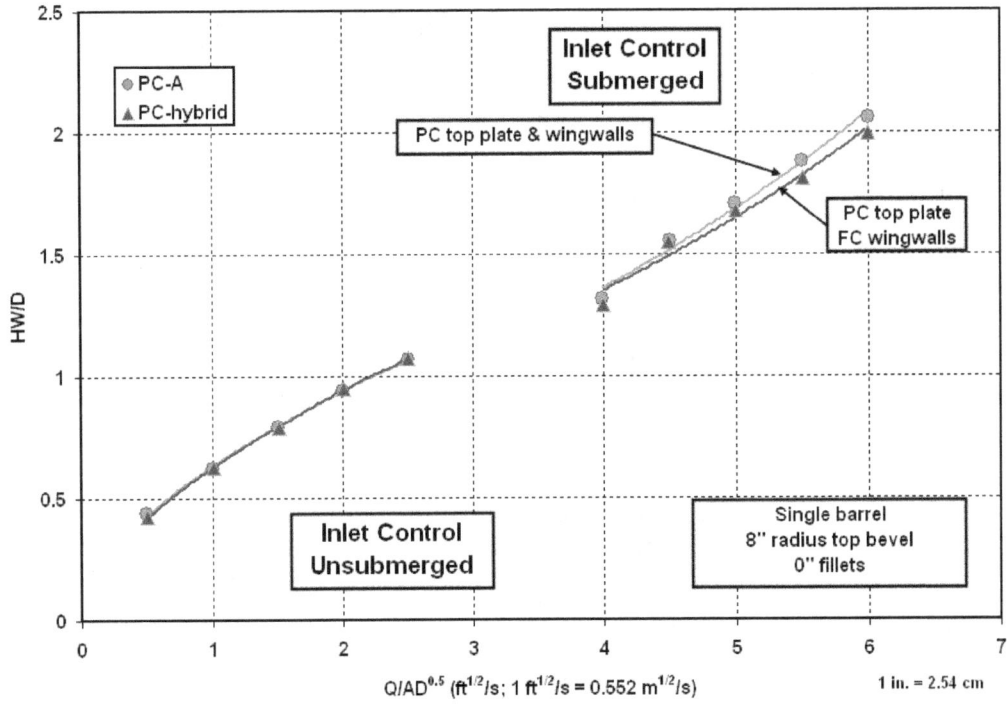

Figure 100. Graph. Inlet control, precast hybrid inlet with no bevel on wingwalls.

The following charts, figures 101 through 122 show the effects of multiple barrels.

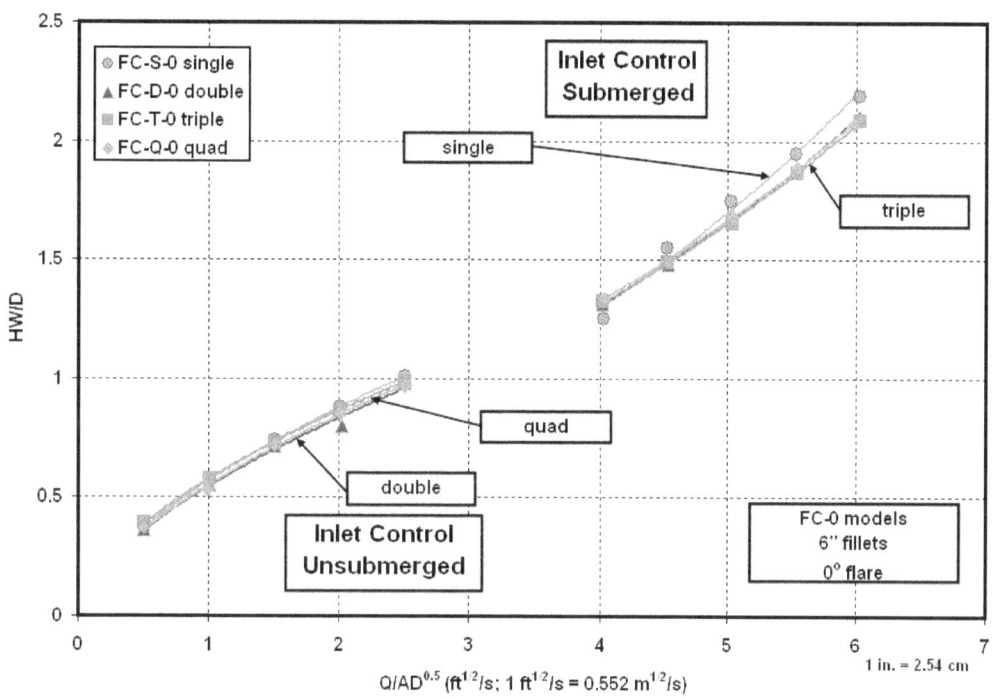

Figure 101. Graph. Inlet control, FC-S-0, FC-D-0, FC-T-0, and FC-Q-0.

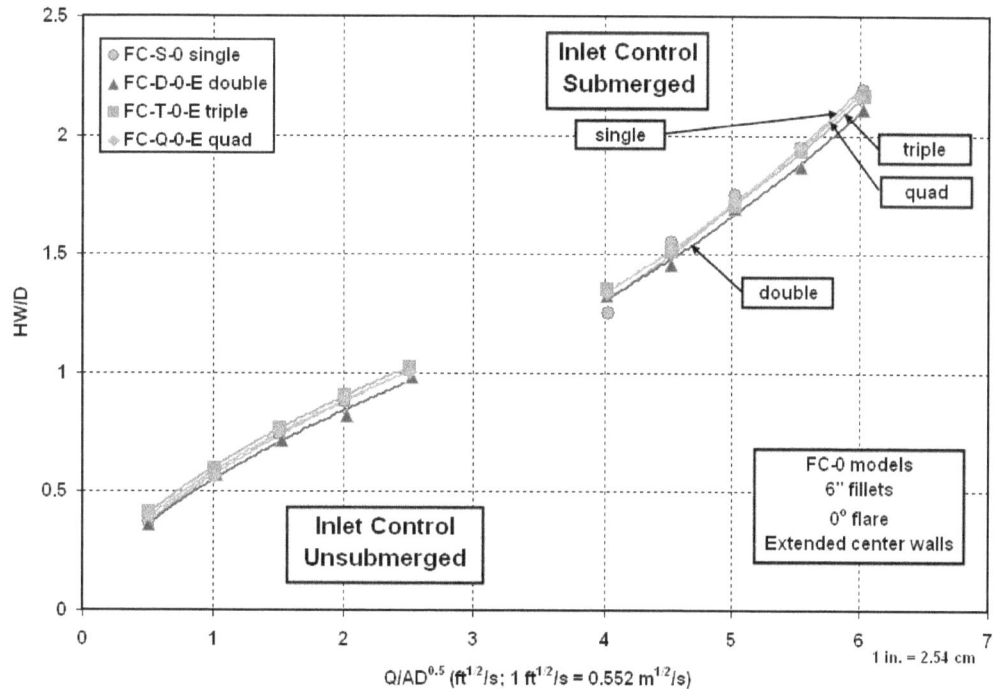

Figure 102. Graph. Inlet control, FC-S-0, FC-D-0-E, FC-T-0-E, and FC-Q-0-E.

103

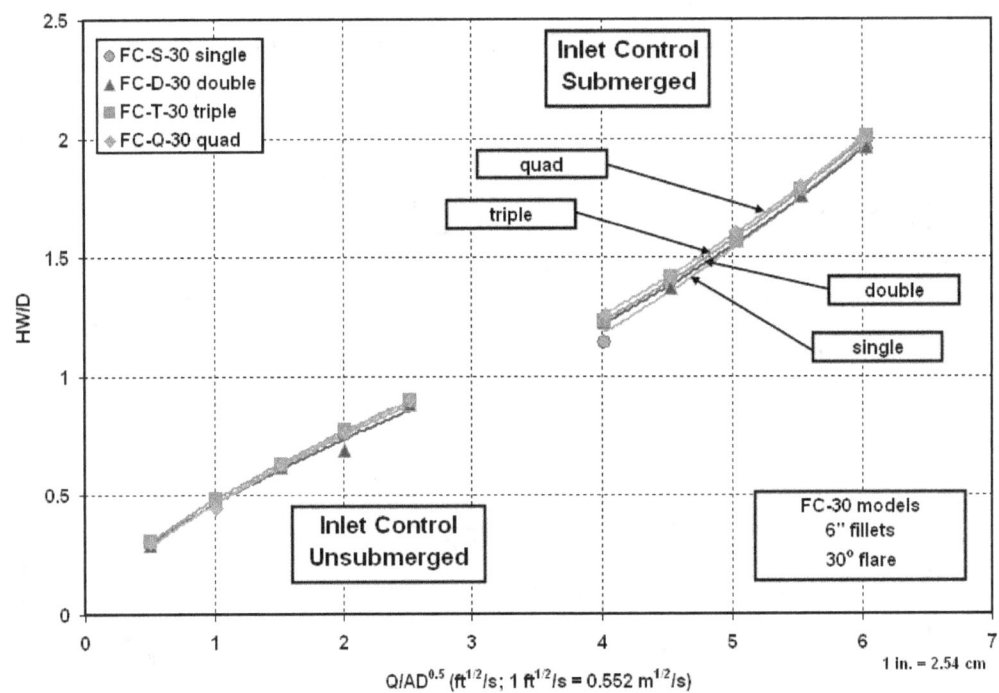

Figure 103. Graph. Inlet control, FC-S-30, FC-D-30, FC-T-30, and FC-Q-30.

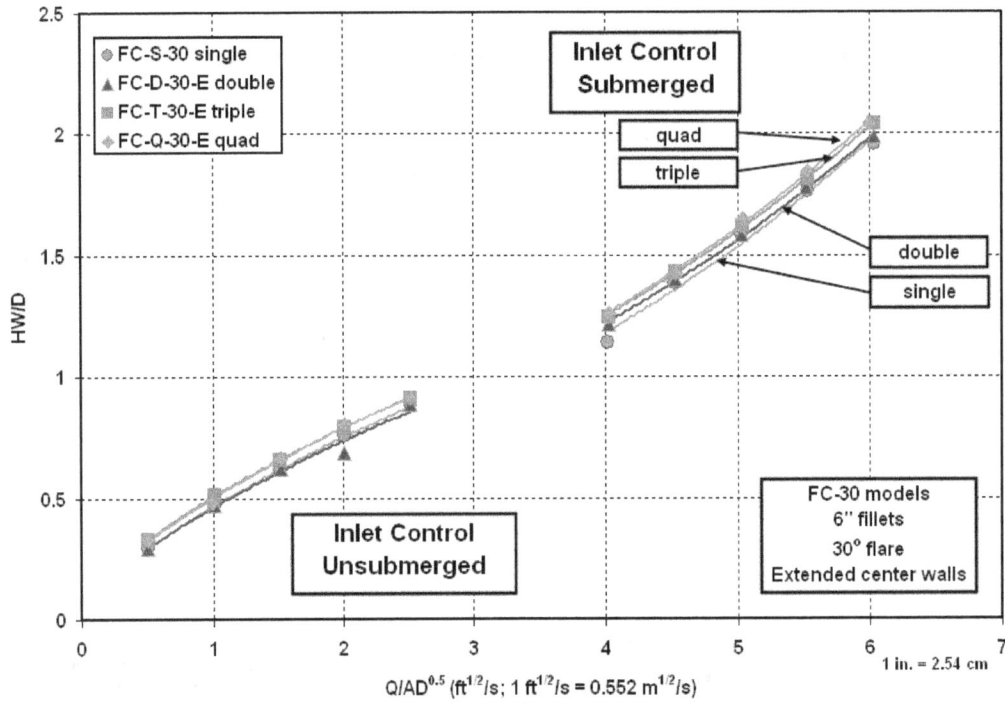

Figure 104. Graph. Inlet control, FC-S-30, FC-D-30-E, FC-T-30-E, and FC-Q-30-E.

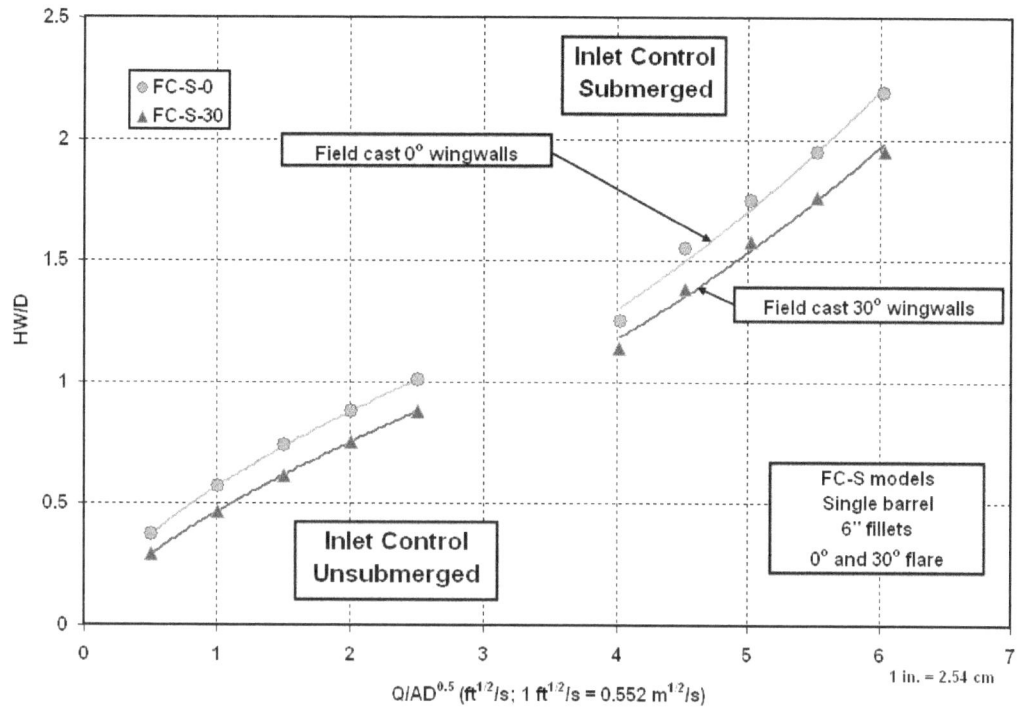

Figure 105. Graph. Inlet control, FC-S-0 and FC-S-30.

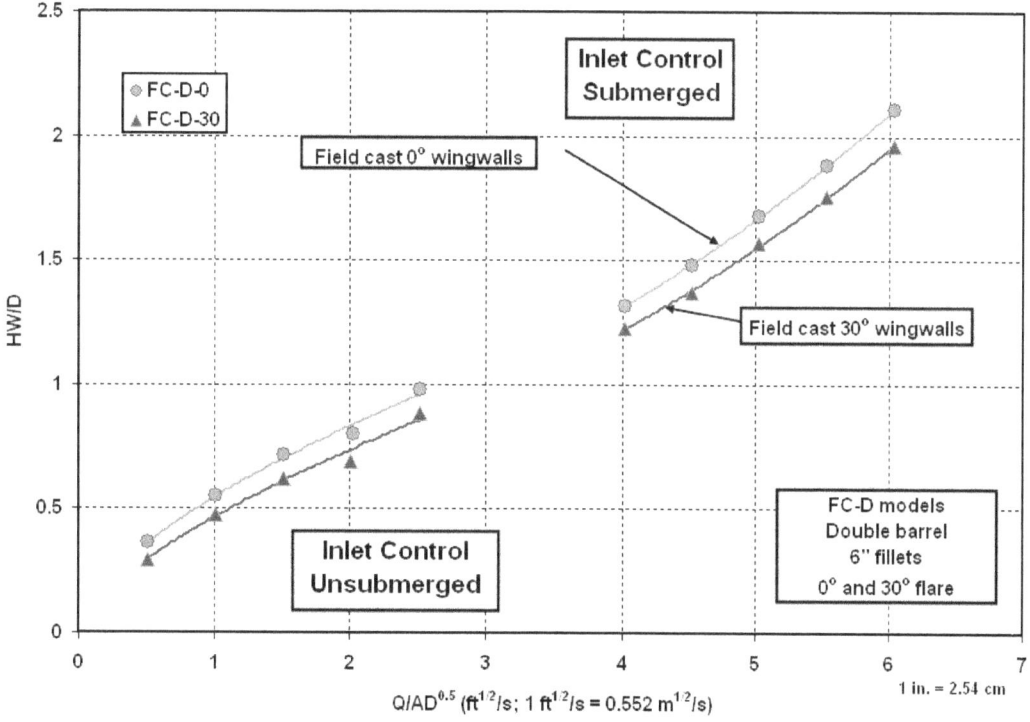

Figure 106. Graph. Inlet control, FC-D-0 and FC-D-30.

105

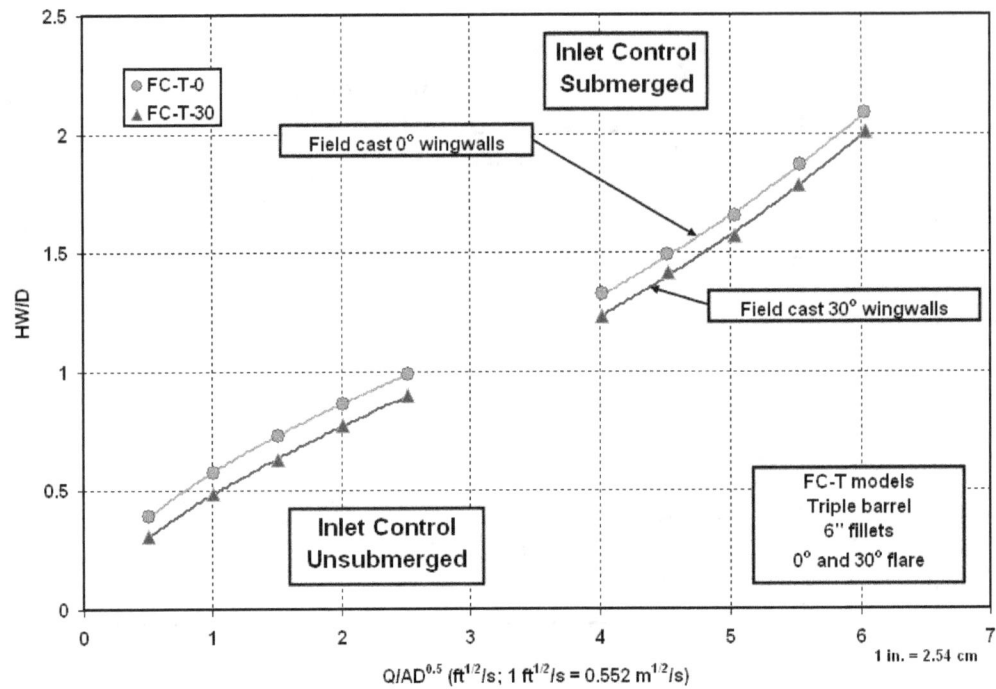

Figure 107. Graph. Inlet control, FC-T-0 and FC-T-30.

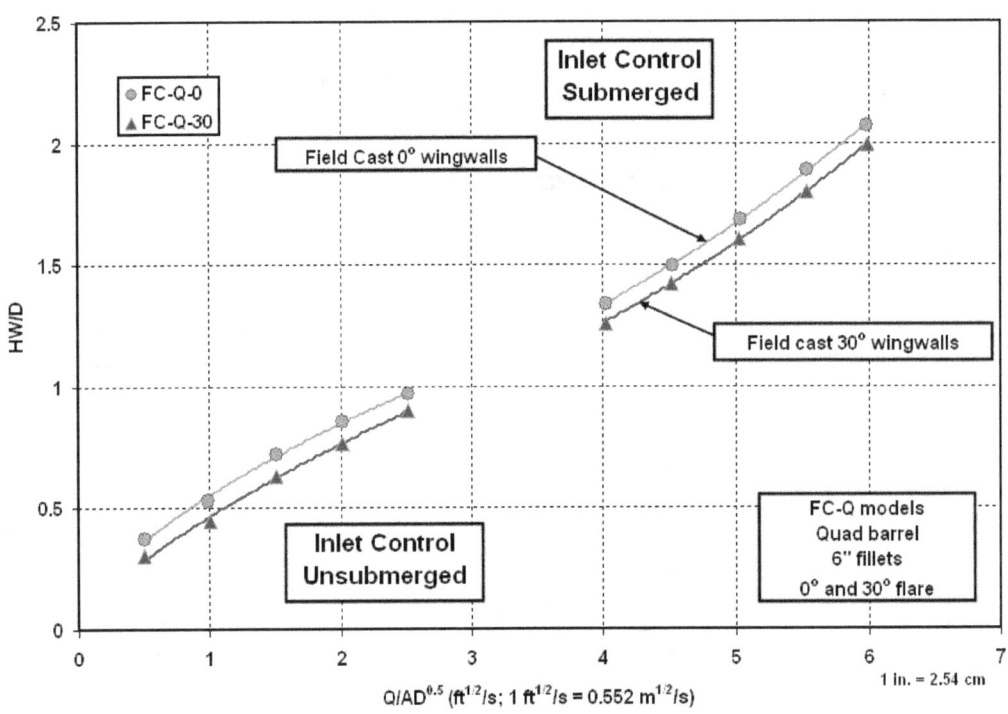

Figure 108. Graph. Inlet control, FC-Q-0 and FC-Q-30.

106

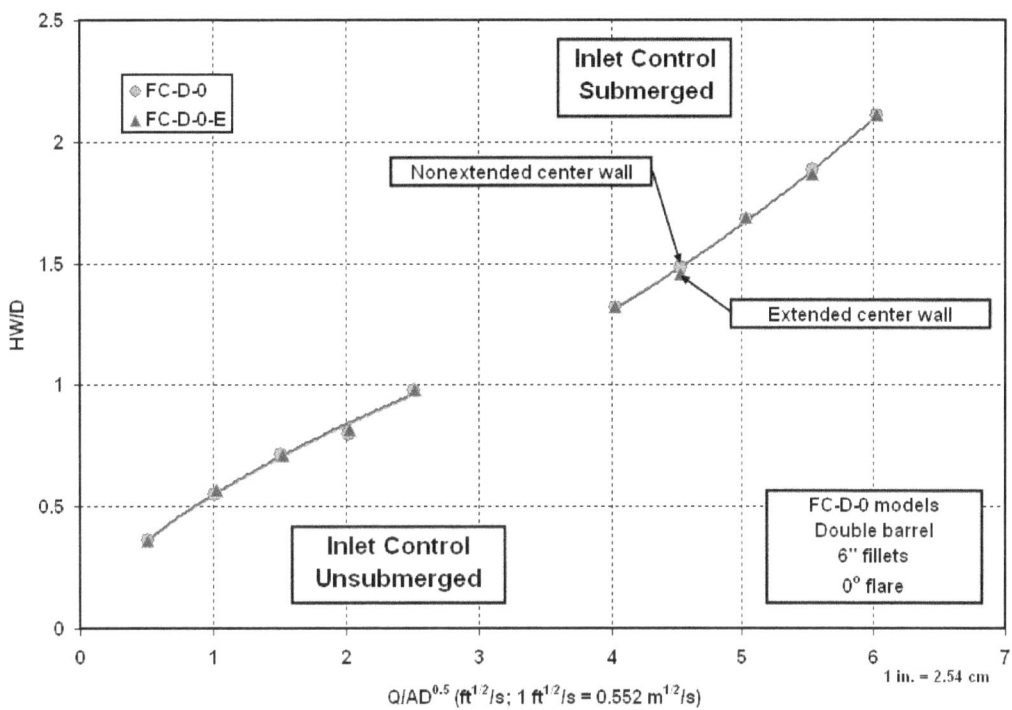

Figure 109. Graph. Inlet control, FC-D-0 and FC-D-0-E.

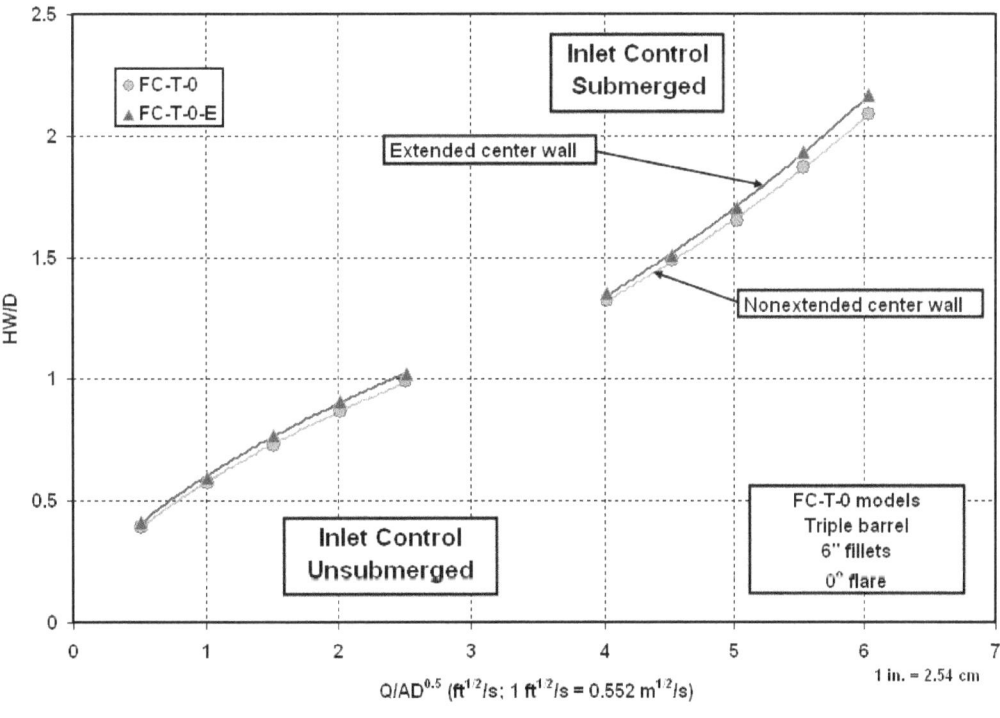

Figure 110. Graph. Inlet control, FC-T-0 and FC-T-0-E.

107

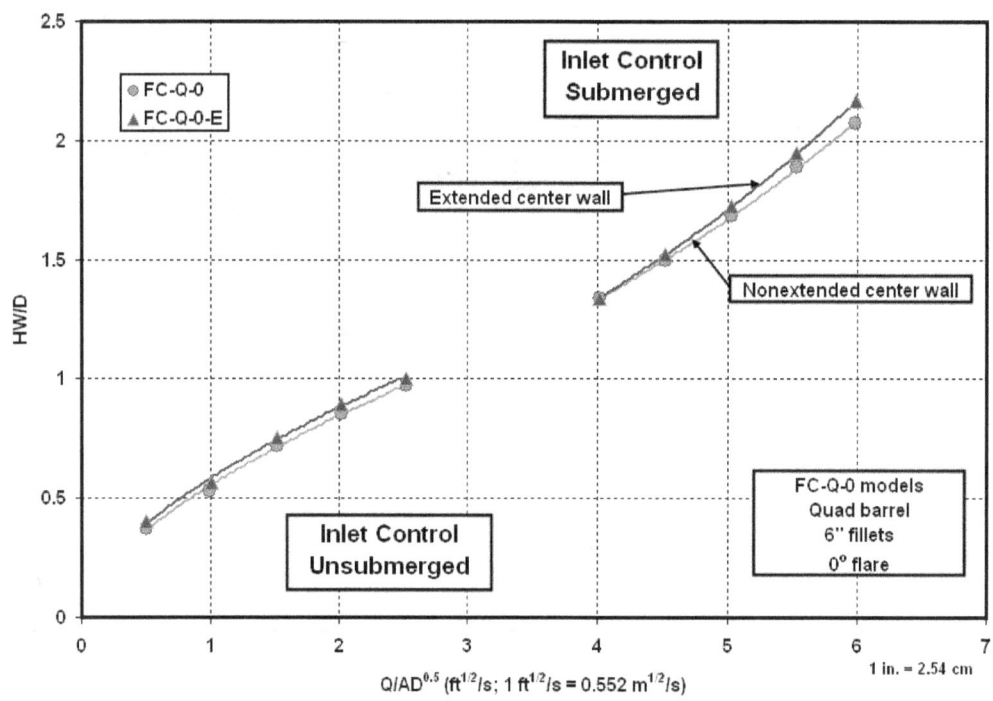

Figure 111. Graph. Inlet control, FC-Q-0 and FC-Q-0-E.

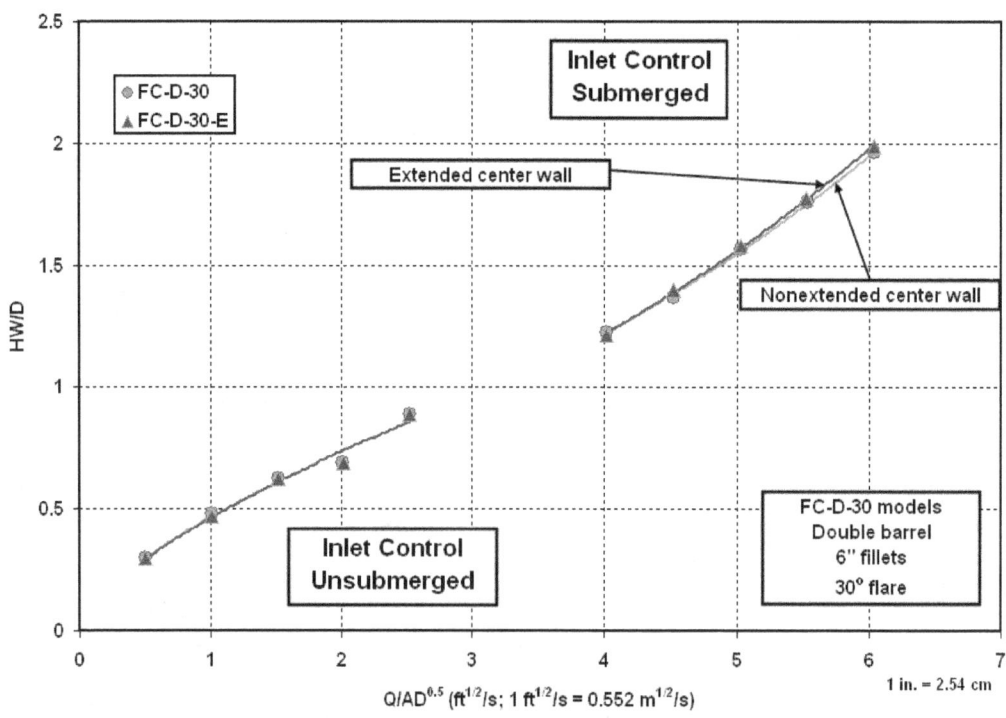

Figure 112. Graph. Inlet control, FC-D-30 and FC-D-30-E.

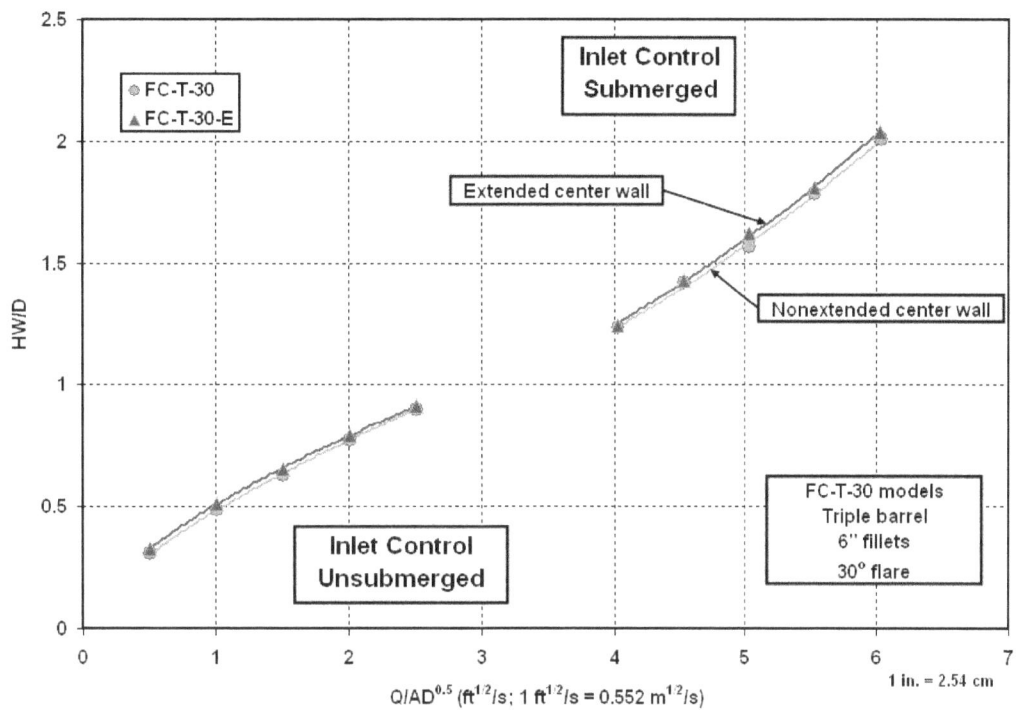

Figure 113. Graph. Inlet control, FC-T-30 and FC-T-30-E.

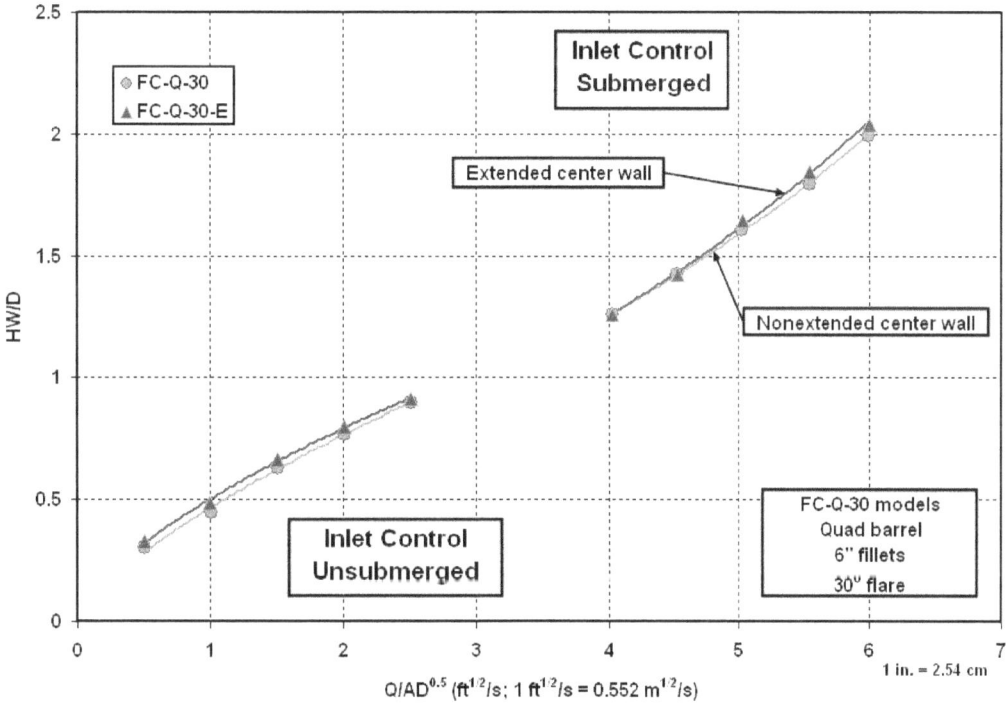

Figure 114. Graph. Inlet control, FC-Q-30 and FC-Q-30-E.

109

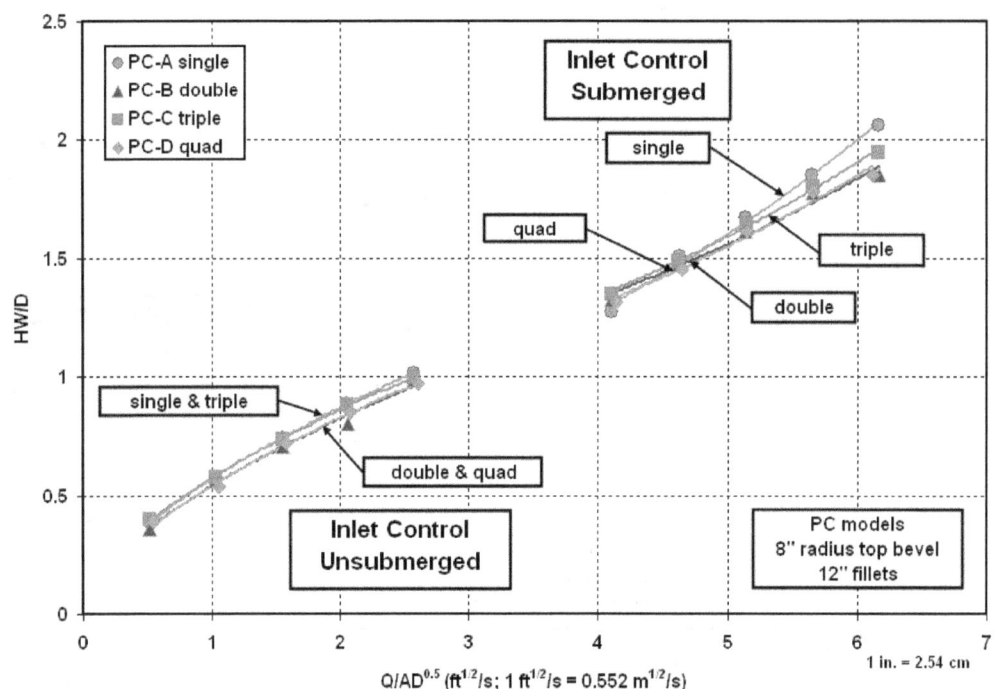

Figure 115. Graph. Inlet control, PC-A, PC-B, PC-C, and PC-D.

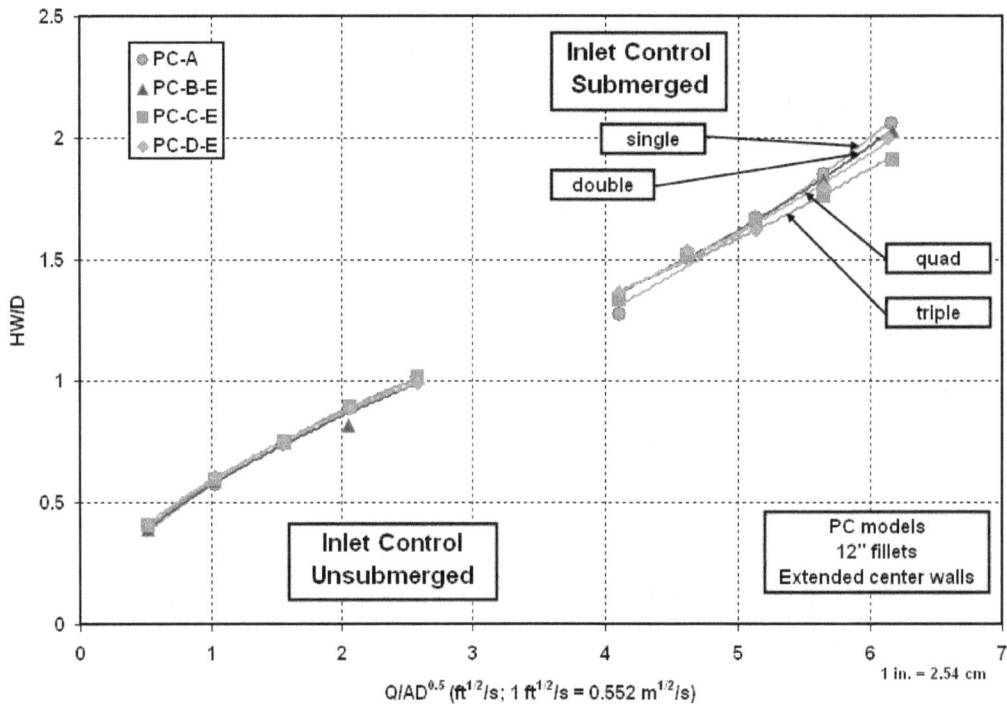

Figure 116. Graph. Inlet control, PC-A, PC-B-E, PC-C-E, and PC-D-E.

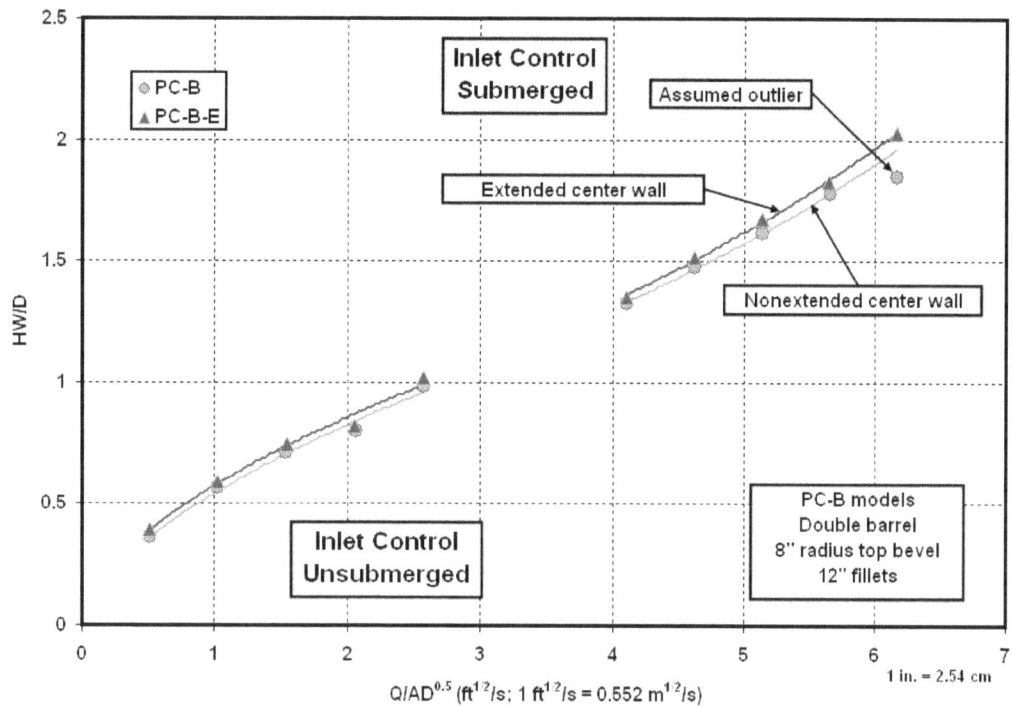

Figure 117. Graph. Inlet control, PC-B and PC-B-E.

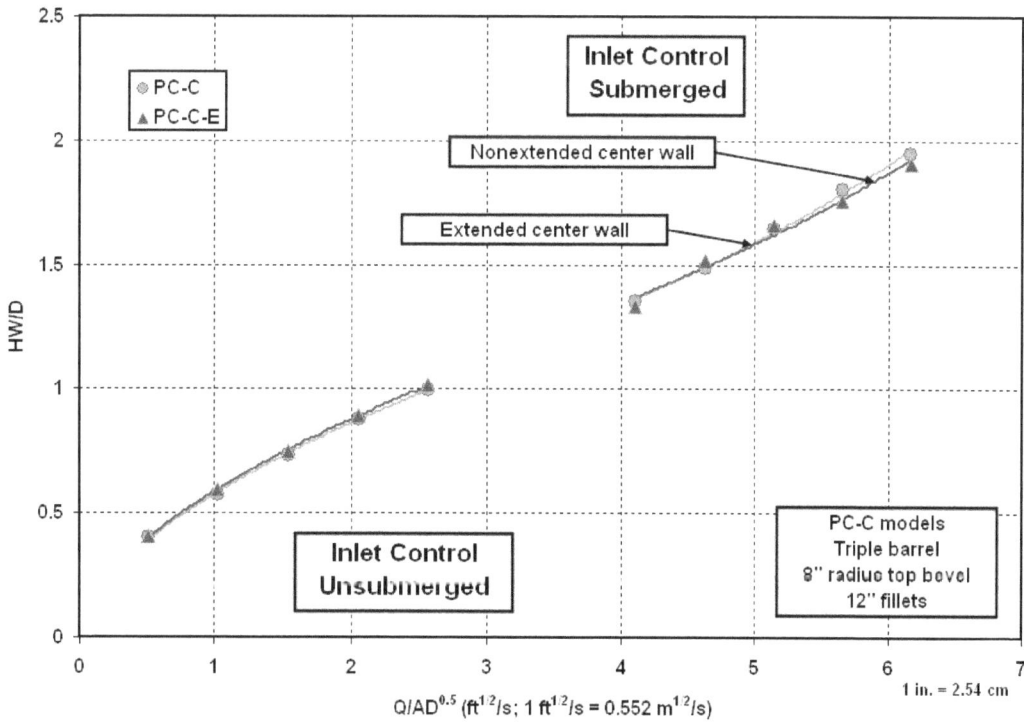

Figure 118. Graph. Inlet control, PC-C and PC-C-E.

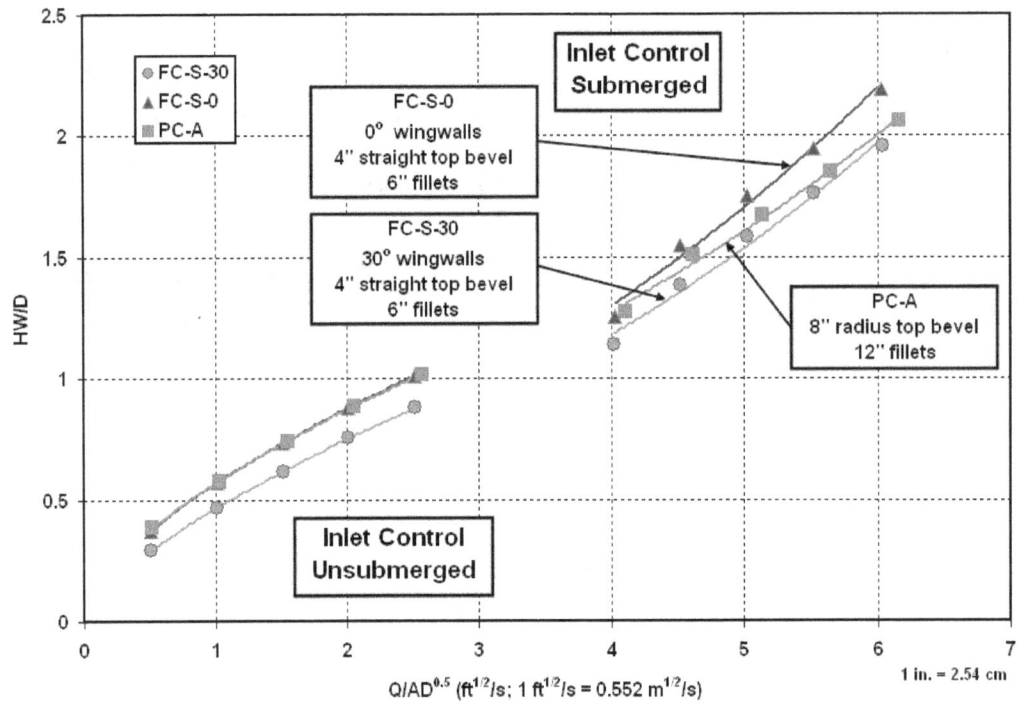

Figure 119. Graph. Inlet control, FC-S-30, FC-S-0, and PC-A.

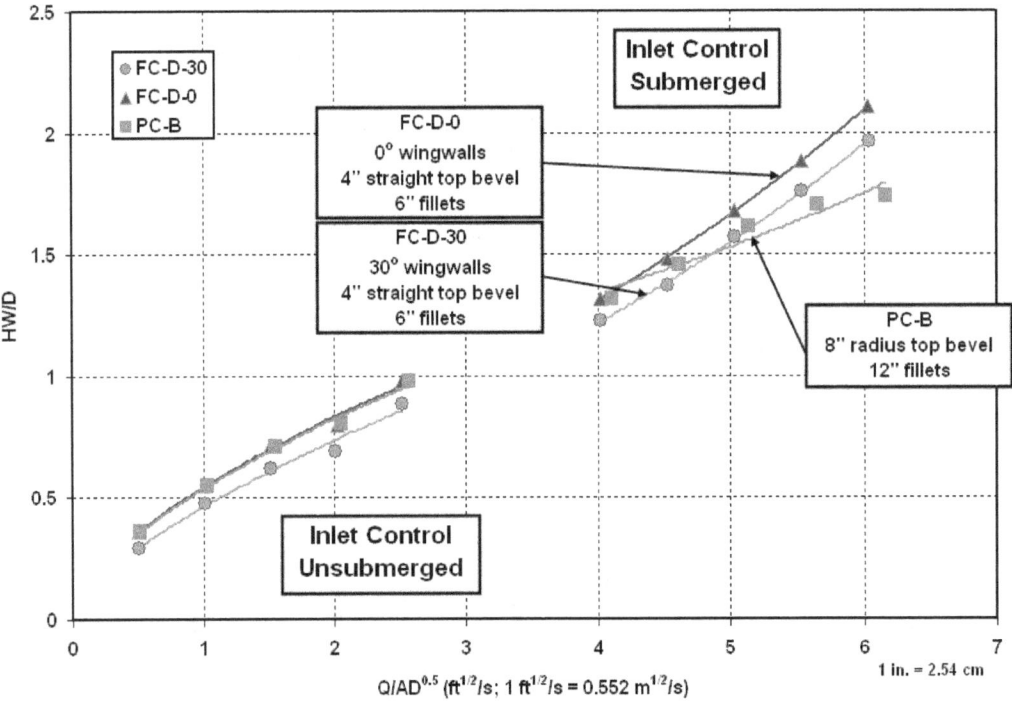

Figure 120. Graph. Inlet control, FC-D-30, FC-D-0, and PC-B.

112

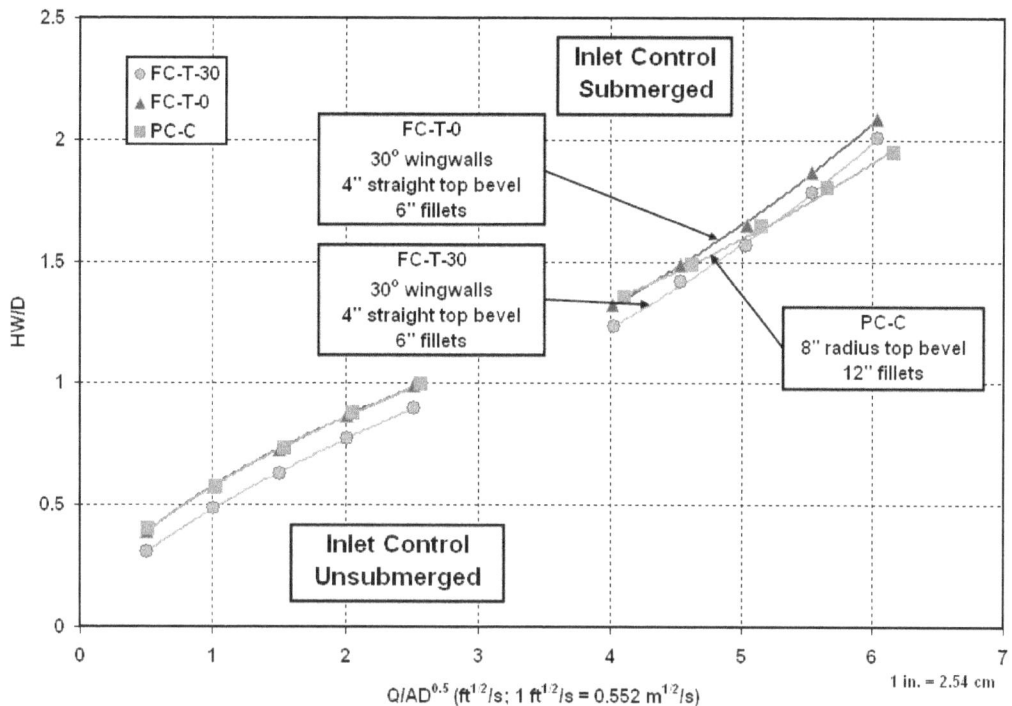

Figure 121. Graph. Inlet control, FC-T-30, FC-T-0, and PC-C.

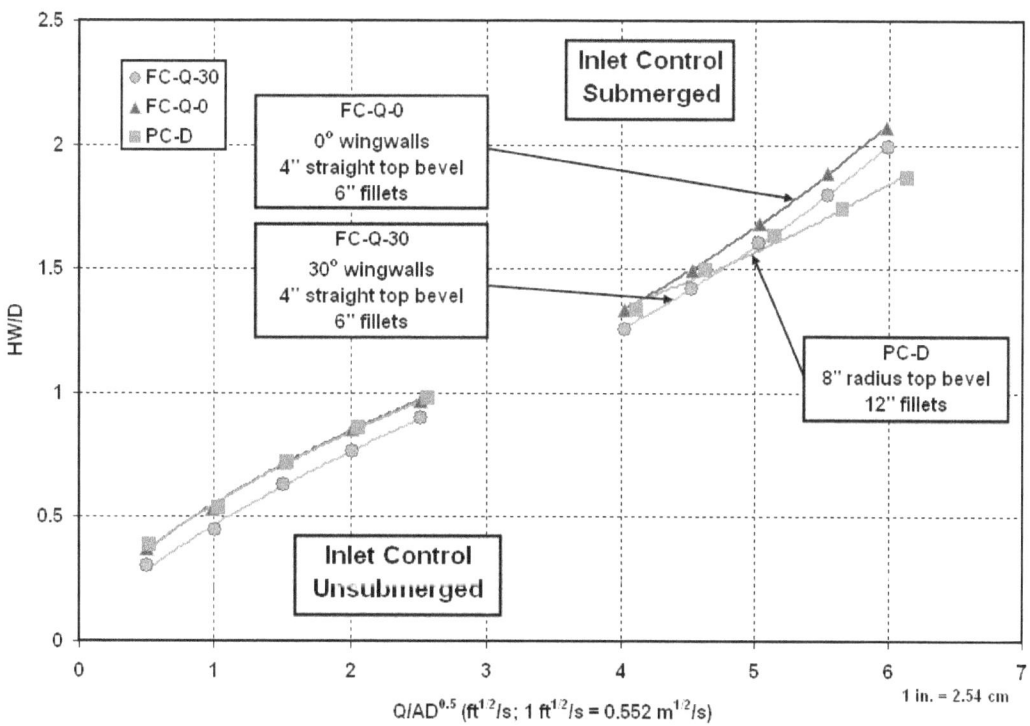

Figure 122. Graph. Inlet control, FC-Q-30, FC-Q-0, and PC-D.

The following charts, figures 123 through 129 show the effects of the span-to-rise ratio.

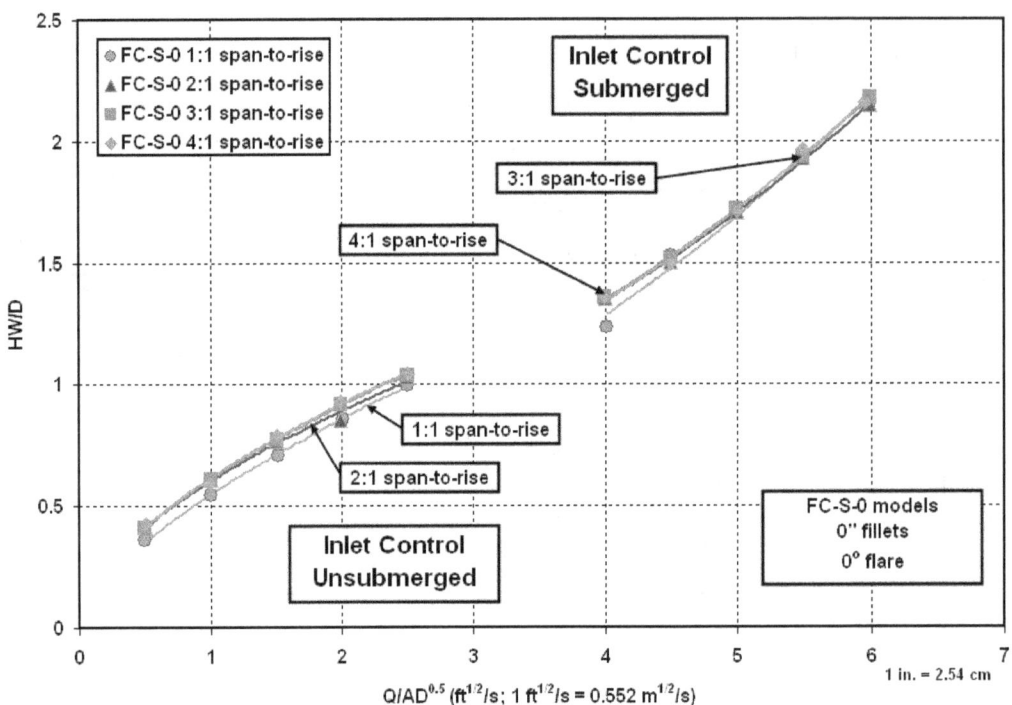

Figure 123. Graph. Inlet control, FC-S-0, various span-to-rise ratios.

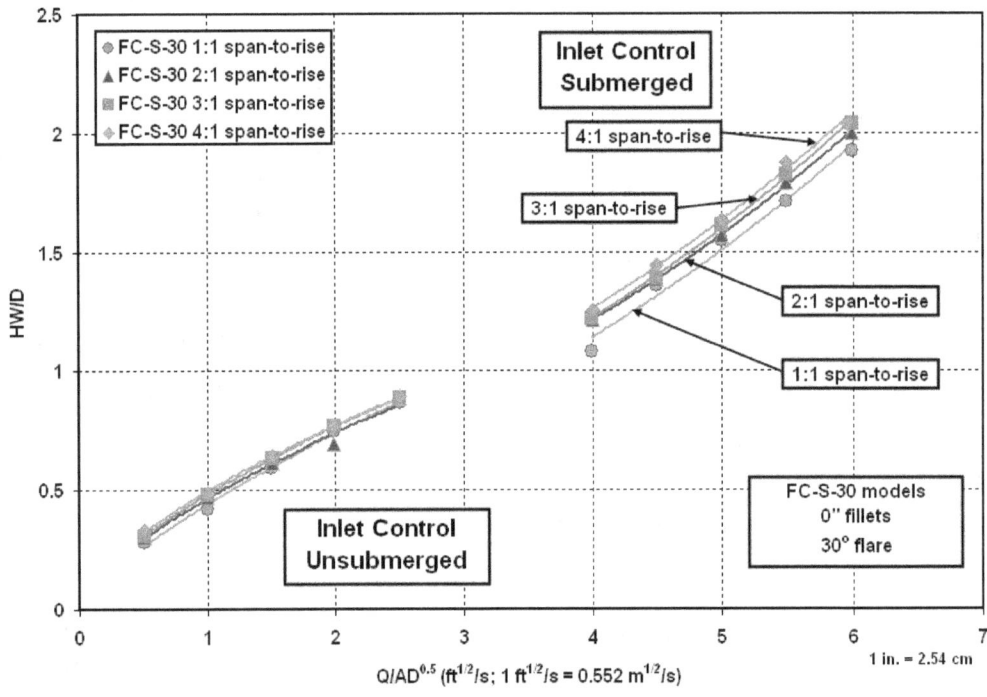

Figure 124. Graph. Inlet control, FC-S-30, various span-to-rise ratios.

114

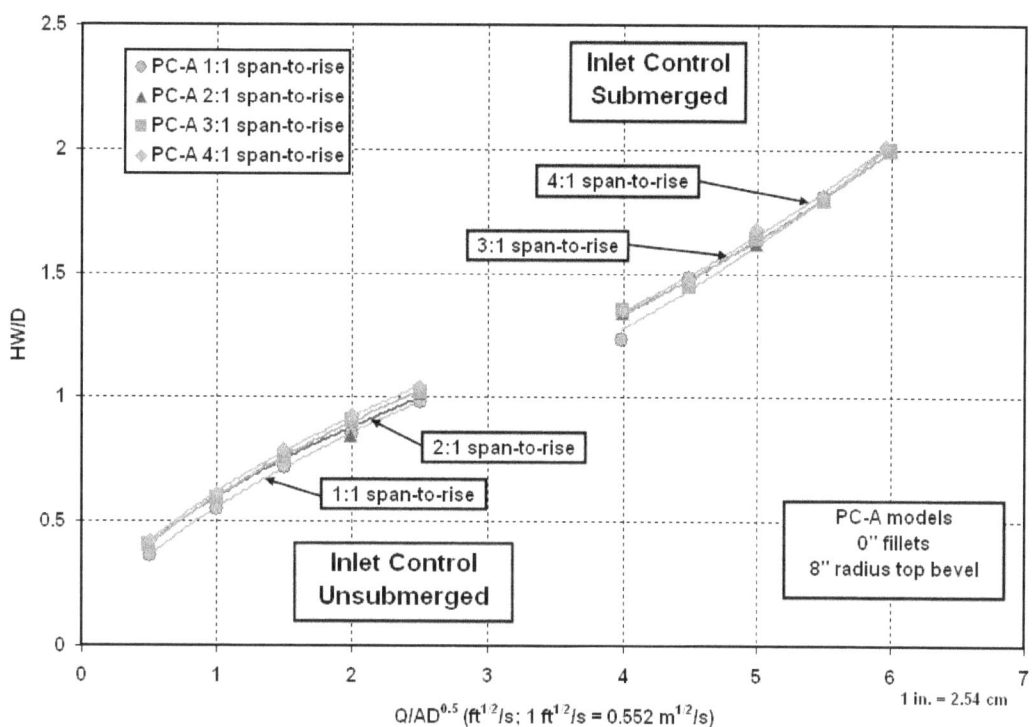

Figure 125. Graph. Inlet control, PC-A, various span-to-rise ratios.

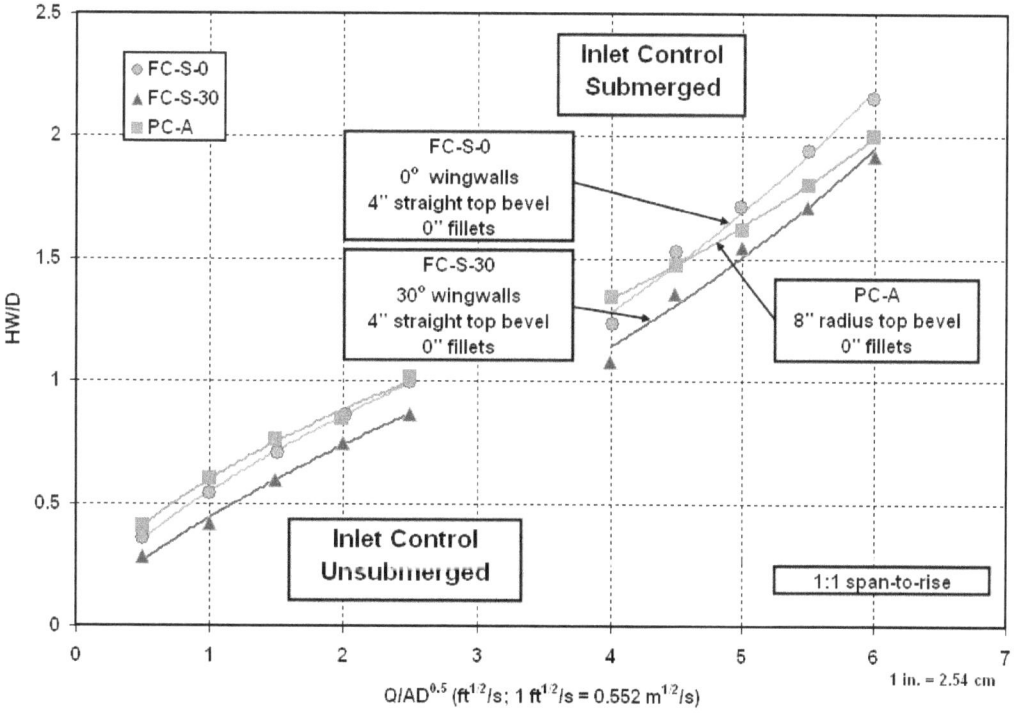

Figure 126. Graph. Inlet control, FC-S-0, FC-S-30, and PC-A, 1:1 span-to-rise ratio.

115

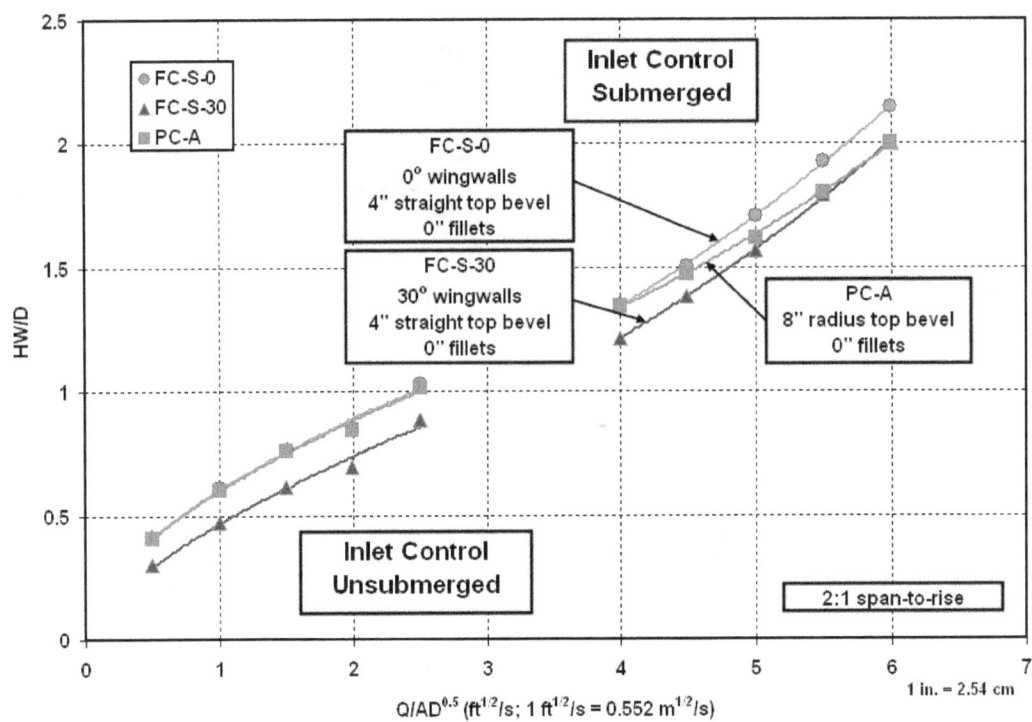

Figure 127. Graph. Inlet control, FC-S-0, FC-S-30, and PC-A, 2:1 span-to-rise ratio.

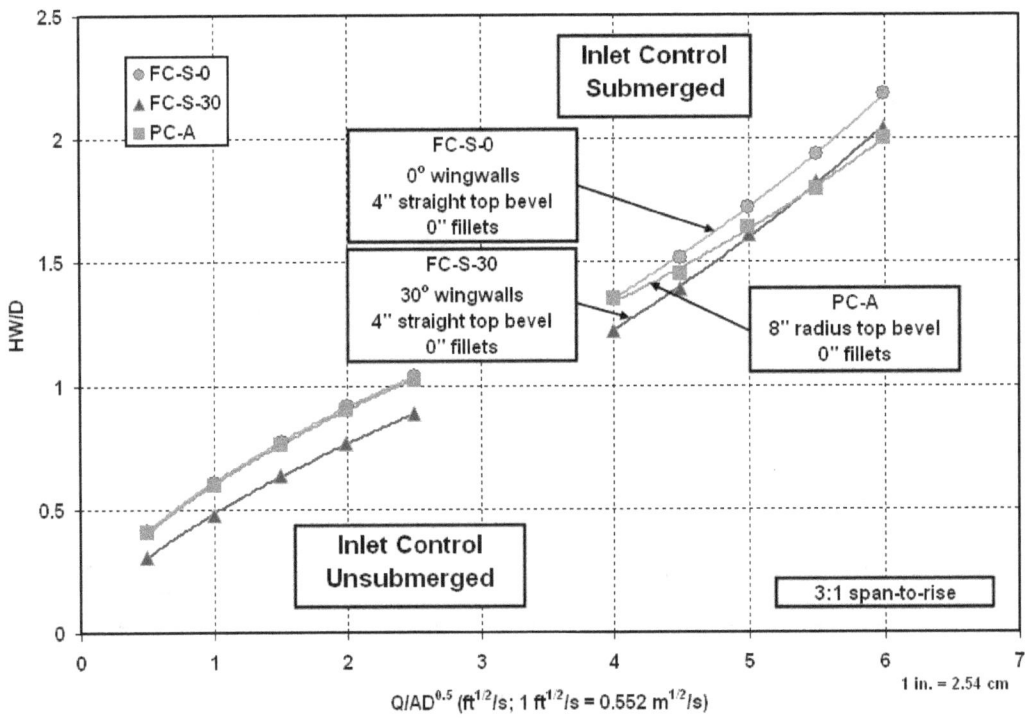

Figure 128. Graph. Inlet control, FC-S-0, FC-S-30, and PC-A, 3:1 span-to-rise ratio.

116

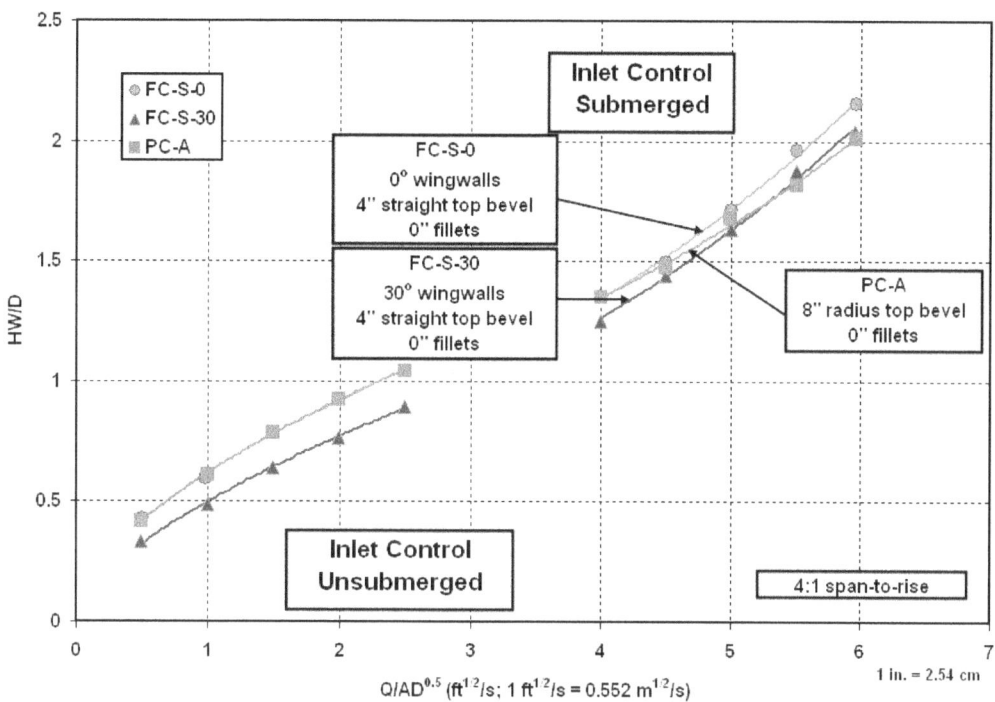

Figure 129. Graph. Inlet control, FC-S-0, FC-S-30, and PC-A, 4:1 span-to-rise ratio.

The following charts, figures 130 and 131, show the effects of skewed headwalls.

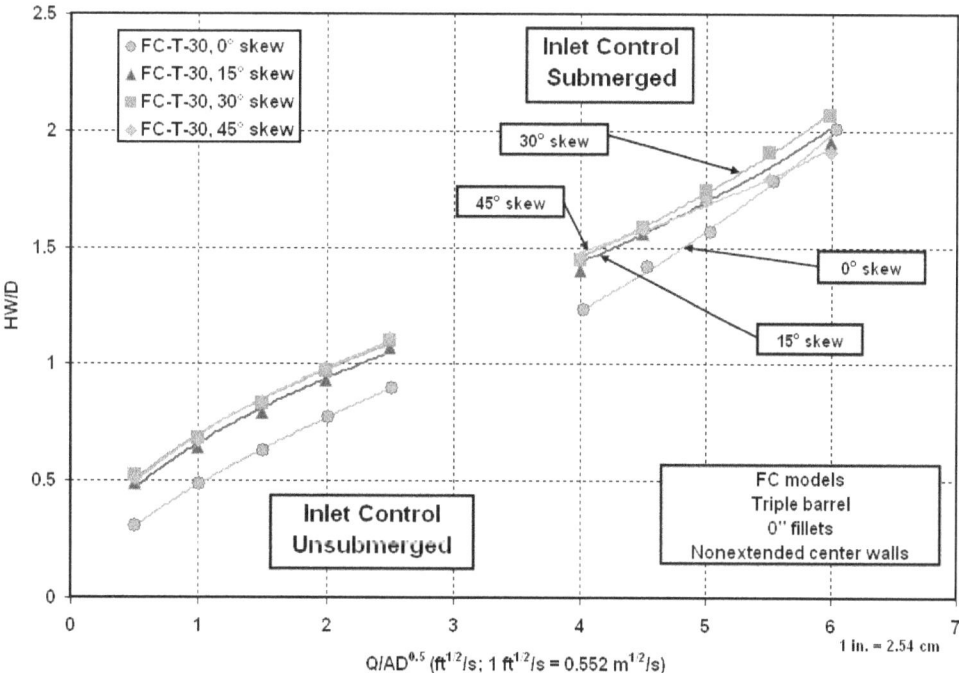

Figure 130. Graph. Inlet control, FC-T-30 at various headwall skews.

117

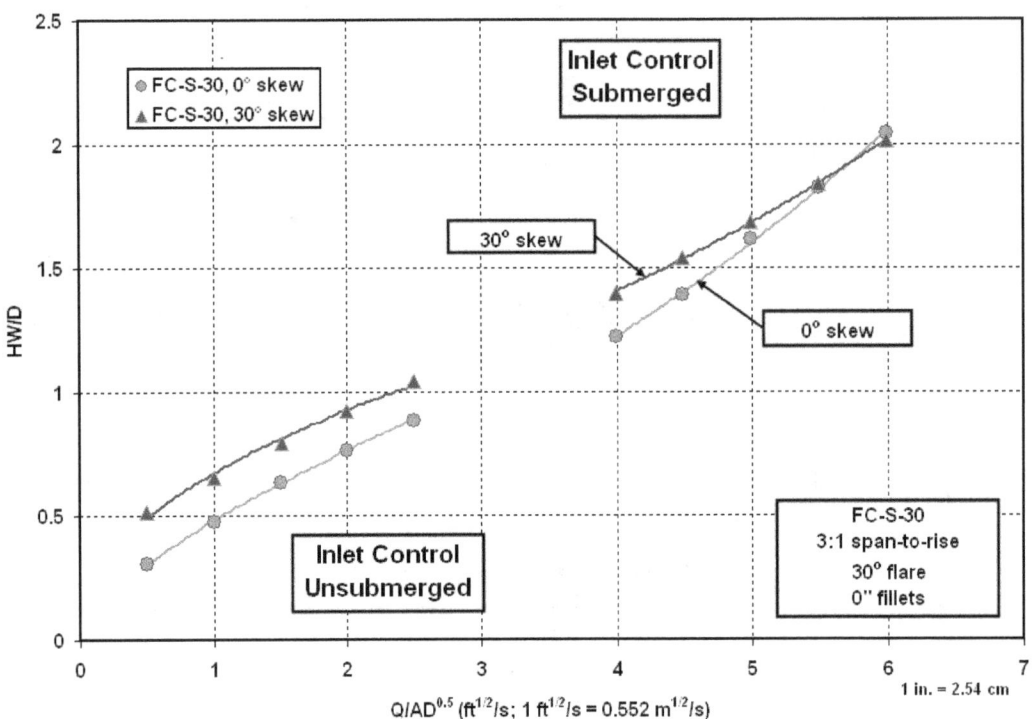

Figure 131. Graph. Inlet control, FC-S-30, 0- and 30-degree skews.

APPENDIX C. SUMMARY OF REGRESSION COEFFICIENTS

Tables 17, 18, and 19 of this appendix summarize the regression coefficients derived from all of the experiments conducted during this study. Table 17 lists the inlet control design coefficients—K, M, c, and Y—that apply to the HDS-5 form 2 unsubmerged and submerged flow equations in figures 14 and 15 in chapter 3.

Table 18 lists the inlet control fifth-order polynomial constants—a, b, c, d, e, and f—that are used by computer programmers to bridge the transition from unsubmerged to submerged flow by coding the two equations as a continuous function. The full fifth-order polynomial used in culvert design is given in figure 132.

$$\frac{HW_i}{D} = a + b\left[\frac{K_u Q}{AD^{0.5}}\right] + c\left[\frac{K_u Q}{AD^{0.5}}\right]^2 + d\left[\frac{K_u Q}{AD^{0.5}}\right]^3 + e\left[\frac{K_u Q}{AD^{0.5}}\right]^4 + f\left[\frac{K_u Q}{AD^{0.5}}\right]^5$$

Figure 132. Equation. Fifth-order polynomial.

K_u is the units conversion constant and equals 1.811 for SI units and 1.0 for customary English units.

Table 19 lists the outlet control loss coefficients K_e and K_o used to compute culvert entrance and culvert outlet losses. The coefficients are commonly expressed as functions of the barrel velocity head, as indicated in the equations in figures 19 and 20 in chapter 3.

The downstream channel velocity for this study was taken as the average velocity in the tailbox using the full width of the tailbox multiplied by the average tailwater depth as the flow area.

Table 17. Inlet control design coefficients, all experiments.

Inlet	Model	Slope (percent)	Fillet (inches)	Barrels	Span: Rise	Skew (degrees)	K	M	c	Y
Bevels and fillets experiments										
1.1	FC-S-0	3.0	0	1	1:1	0	0.55	0.64	0.0453	0.54
1.2	FC-S-0	3.0	6	1	1:1	0	0.57	0.62	0.0448	0.56
1.3	FC-S-0	3.0	12	1	1:1	0	0.58	0.61	0.0447	0.54
1.4	PC-A	3.0	0	1	1:1	0	0.56	0.63	0.0371	0.67
1.5	PC-A	3.0	6	1	1:1	0	0.57	0.62	0.0371	0.67
1.6	PC-A	3.0	12	1	1:1	0	0.57	0.61	0.0361	0.68
1.7	PC-A	3.0	6	1	2:1	0	0.60	0.56	0.0329	0.79
1.8	PC-A	3.0	12	1	2:1	0	0.60	0.55	0.0316	0.81
1.9	PC-A hybrid	3.0	0	1	1:1	0	0.63	0.58	0.0331	0.80
1.10	FC-S-0 hybrid	3.0	0	1	1:1	0	0.62	0.59	0.0445	0.66
Multiple barrel experiments										
2.1	FC-S-0	3.0	6	1	1:1	0	0.57	0.62	0.0448	0.56
2.2	FC-S-30	3.0	6	1	1:1	0	0.47	0.69	0.0394	0.53
2.3	FC-D-0	3.0	6	2	1:1	0	0.55	0.61	0.0391	0.66
2.4	FC-D-0-E	3.0	6	2	1:1	0	0.55	0.61	0.0394	0.65
2.5	FC-D-30	3.0	6	2	1:1	0	0.47	0.66	0.0366	0.61
2.6	FC-D-30-E	3.0	6	2	1:1	0	0.46	0.67	0.0381	0.59
2.7	FC-T-0	3.0	6	3	1:1	0	0.58	0.58	0.0377	0.69
2.8	FC-T-0-E	3.0	6	3	1:1	0	0.60	0.57	0.0405	0.67
2.9	FC-T-30	3.0	6	3	1:1	0	0.48	0.67	0.0379	0.60
2.10	FC-T-30-E	3.0	6	3	1:1	0	0.51	0.64	0.0405	0.60
2.11	FC-Q-0	3.0	6	4	1:1	0	0.55	0.61	0.0377	0.71
2.12	FC-Q-0-E	3.0	6	4	1:1	0	0.58	0.59	0.0418	0.64
2.13	FC-Q-30	3.0	6	4	1:1	0	0.47	0.71	0.0372	0.64
2.14	FC-Q-30-E	3.0	6	4	1:1	0	0.50	0.65	0.0398	0.60
2.15	PC-A	3.0	12	1	1:1	0	0.57	0.61	0.0361	0.68
2.16	PC-B	3.0	12	2	1:1	0	0.54	0.61	0.0253	0.91
2.17	PC-B-E	3.0	12	2	1:1	0	0.57	0.58	0.0315	0.81
2.18	PC-C	3.0	12	3	1:1	0	0.57	0.57	0.0219	0.98
2.19	PC-C-E	3.0	12	3	1:1	0	0.59	0.58	0.0263	0.91
2.20	PC-D	3.0	12	4	1:1	0	0.55	0.61	0.0250	0.92
2.21	PC-D-E	3.0	12	4	1:1	0	0.60	0.54	0.0296	0.85

1 inch = 2.54 cm

Table 17. Inlet control design coefficients, all experiments—*Continued.*

Inlet	Model	Slope (percent)	Fillet (inches)	Barrels	Span: Rise	Skew (degrees)	K	M	c	Y
Span-to-rise experiments										
3.1	FC-S-0	3.0	0	1	1:1	0	0.55	0.64	0.0453	0.54
3.2	FC-S-0	3.0	0	1	2:1	0	0.61	0.56	0.0404	0.68
3.3	FC-S-0	3.0	0	1	3:1	0	0.61	0.58	0.0413	0.67
3.4	FC-S-0	3.0	0	1	4:1	0	0.62	0.57	0.0421	0.65
3.5	FC-S-30	3.0	0	1	1:1	0	0.44	0.74	0.0403	0.48
3.6	FC-S-30	3.0	0	1	2:1	0	0.47	0.66	0.0397	0.56
3.7	FC-S-30	3.0	0	1	3:1	0	0.48	0.66	0.0414	0.54
3.8	FC-S-30	3.0	0	1	4:1	0	0.50	0.63	0.0410	0.59
3.9	PC-A	3.0	0	1	1:1	0	0.56	0.63	0.0371	0.67
3.10	PC-A	3.0	0	1	2:1	0	0.60	0.56	0.0329	0.79
3.11	PC-A	3.0	0	1	3:1	0	0.60	0.58	0.0331	0.79
3.12	PC-A	3.0	0	1	4:1	0	0.62	0.57	0.0340	0.79
Skewed headwall experiments										
4.01	FC-T-0	3.0	6	3	1:1	0	0.58	0.58	0.0377	0.69
4.02	FC-T-30	3.0	6	3	1:1	0	0.48	0.67	0.0369	0.62
4.03	FC-T-30	3.0	0	3	1:1	15	0.66	0.51	0.0289	0.95
4.04	FC-T-30	3.0	0	3	1:1	30	0.70	0.48	0.0312	0.94
4.05	FC-T-30	3.0	0	3	1:1	45	0.69	0.50	0.0224	1.10
4.06	FC-S-30	3.0	0	1	3:1	30	0.68	0.46	0.0306	0.89

1 inch = 2.54 cm

Table 18. Inlet control fifth-order polynomial coefficients, all experiments.

Inlet	Model	Slope (percent)	Fillet (inches)	Barrels	Span: Rise	Skew (degrees)	a	b	c	d	e	f
Bevels and fillets experiments												
1.1	FC-S-0	3.0	0	1	1:1	0	0.211536	0.224341	0.208370	-0.12199	0.024676	-0.00161
1.2	FC-S-0	3.0	6	1	1:1	0	0.224471	0.247312	0.186514	-0.11350	0.023189	-0.00151
1.3	FC-S-0	3.0	12	1	1:1	0	0.245464	0.218175	0.210202	-0.12126	0.024063	-0.00153
1.4	PC-A	3.0	0	1	1:1	0	0.194217	0.310678	0.109365	-0.07741	0.016183	-0.00106
1.5	PC-A	3.0	6	1	1:1	0	0.203074	0.313529	0.107677	-0.07685	0.016046	-0.00104
1.6	PC-A	3.0	12	1	1:1	0	0.210576	0.314554	0.101951	-0.07265	0.014939	-0.00095
1.7	PC-A	3.0	6	1	2:1	0	0.186778	0.482282	-0.07044	-0.00309	0.003186	-0.00026
1.8	PC-A	3.0	12	1	2:1	0	0.189232	0.496842	0.090840	0.005876	0.001525	-0.00015
1.9	PC-A hybrid	3.0	0	1	1:1	0	0.240477	0.307282	0.167334	-0.11478	0.024511	-0.00168
1.10	FC-S-0hybrid	3.0	0	1	1:1	0	0.324959	0.032735	0.428664	-0.22300	0.044585	-0.00300
Multiple barrel experiments												
2.1	FC-S-0	3.0	6	1	1:1	0	0.224471	0.247312	0.186514	-0.11350	0.023189	-0.00151
2.2	FC-S-30	3.0	6	1	1:1	0	0.148953	0.250437	0.129334	-0.08043	0.016465	-0.00106
2.3	FC-D-0	3.0	6	2	1:1	0	0.154694	0.436581	-0.03432	-0.01750	0.006033	-0.00044
2.4	FC-D-0-E	3.0	6	2	1:1	0	0.168455	0.406534	-0.00092	-0.03180	0.008541	-0.00060
2.5	FC-D-30	3.0	6	2	1:1	0	0.101044	0.419241	-0.05507	-0.00269	0.002885	-0.00023
2.6	FC-D-30-E	3.0	6	2	1:1	0	0.103678	0.402060	-0.03596	-0.01110	0.004465	-0.00033
2.7	FC-T-0	3.0	6	3	1:1	0	0.185570	0.425924	-0.01513	-0.02737	0.007816	-0.00055
2.8	FC-T-0-E	3.0	6	3	1:1	0	0.220522	0.370427	0.057110	-0.06003	0.013803	-0.00093
2.9	FC-T-30	3.0	6	3	1:1	0	0.132482	0.340923	0.042681	-0.04425	0.010089	-0.00067
2.10	FC-T-30-E	3.0	6	3	1:1	0	0.146450	0.369886	0.018327	-0.03716	0.009256	-0.00064
2.11	FC-Q-0	3.0	6	4	1:1	0	0.154223	0.448350	-0.04069	-0.01406	0.005238	-0.00039
2.12	FC-Q-0-E	3.0	6	4	1:1	0	0.202572	0.369396	0.056302	-0.05916	0.013704	-0.00093
2.13	FC-Q-30	3.0	6	4	1:1	0	0.108394	0.355482	0.027300	-0.03433	0.007929	-0.00052
2.14	FC-Q-30-E	3.0	6	4	1:1	0	0.144926	0.350135	0.041407	-0.04611	0.010763	-0.00073

1 inch = 2.54 cm

122

Table 18. Inlet control fifth-order polynomial coefficients, all experiments—*Continued.*

Inlet	Model	Slope (percent)	Fillet (inches)	Barrels	Span: Rise	Skew (degrees)	a	b	c	d	e	f
2.15	PC-A	3.0	12	1	1:1	0	0.210576	0.314554	0.101951	-0.07265	0.014939	-0.00095
2.16	PC-B	3.0	12	1	1:1	0	0.099284	0.607914	-0.22159	0.066777	-0.00985	0.000567
2.17	PC-B-E	3.0	12	2	1:1	0	0.149381	0.540449	-0.14226	0.029009	-0.00258	0.000099
2.18	PC-C	3.0	12	2	1:1	0	0.147326	0.545578	-0.14618	0.032208	-0.00349	0.000151
2.19	PC-C-E	3.0	12	3	1:1	0	0.153250	0.555777	-0.15207	0.035094	-0.00419	0.000219
2.20	PC-D	3.0	12	3	1:1	0	0.107009	0.600450	-0.20745	0.060926	-0.00892	0.000516
2.21	PC-D-E	3.0	12	4	1:1	0	0.158898	0.585920	-0.19181	0.048483	-0.00586	0.000294
Span-to-rise experiments												
3.1	FC-S-0	3.0	0	1	1:1	0	0.211536	0.224341	0.208370	-0.12199	0.024676	-0.00161
3.2	FC-S-0	3.0	0	1	2:1	0	0.207152	0.432143	-0.01180	-0.03235	0.009224	-0.00067
3.3	FC-S-0	3.0	0	1	3:1	0	0.227185	0.361402	0.077047	-0.06998	0.015731	-0.00106
3.4	FC-S-0	3.0	0	1	4:1	0	0.246621	0.315820	0.126140	-0.09144	0.019626	-0.00131
3.5	FC-S-30	3.0	0	1	1:1	0	0.163450	0.127103	0.256193	-0.13163	0.025211	-0.00160
3.6	FC-S-30	3.0	0	1	2:1	0	0.114704	0.376884	-0.00741	-0.02427	0.006977	-0.00050
3.7	FC-S-30	3.0	0	1	3:1	0	0.141479	0.321334	0.064086	-0.05613	0.012732	-0.00086
3.8	FC-S-30	3.0	0	1	4:1	0	0.23000	0.117000	0.241000	-0.12600	0.025000	-0.00164
3.9	PC-A	3.0	0	1	1:1	0	0.194217	0.310678	0.109365	-0.07741	0.016183	-0.00106
3.10	PC-A	3.0	0	1	2:1	0	0.154724	0.592825	-0.19007	0.048610	-0.00636	0.000370
3.11	PC-A	3.0	0	1	3:1	0	0.200747	0.424467	0.000339	-0.03291	0.008306	-0.00057
3.12	PC-A	3.0	0	1	4:1	0	0.220017	0.404031	0.029449	-0.04658	0.010869	-0.00073
Skewed headwall experiments												
4.01	FC-T-0	3.0	6	3	1:1	0	0.183389	0.409080	0.005613	-0.03583	0.009284	-0.00065
4.02	FC-T-30	3.0	6	3	1:1	0	0.128409	0.355047	0.026694	-0.03750	0.008897	-0.00060
4.03	FC-T-30	3.0	0	3	1:1	15	0.182031	0.686256	-0.27704	0.082113	-0.01167	0.000654
4.04	FC-T-30	3.0	0	3	1:1	30	0.225978	0.658427	-0.24032	0.063449	-0.00797	0.000409
4.05	FC-T-30	3.0	0	3	1:1	45	0.187459	0.743672	-0.32364	0.103601	-0.01612	0.000958
4.06	FC-S-30	3.0	0	1	3:1	30	0.213123	0.685460	-0.29816	0.088583	-0.01224	0.000663

1 inch = 2.54 cm

Table 19. Outlet control design coefficients, all experiments.

Inlet	Model	Slope (percent)	Fillet (inches)	Barrels	Span: Rise	Skew (degrees)	Outlet loss K_o unsubmerged	Outlet loss K_o submerged	Entrance loss K_e unsubmerged	Entrance loss K_e submerged
Bevels and fillets experiments										
1.1	FC-S-0	3.0	0	1	1:1	0	NA	1.22	NA	0.45
1.2	FC-S-0	3.0	6	1	1:1	0	NA	1.25	NA	0.47
1.3	FC-S-0	3.0	12	1	1:1	0	NA	1.17	NA	0.64
1.4	PC-A	3.0	0	1	1:1	0	NA	1.22	NA	0.25
1.5	PC-A	3.0	6	1	1:1	0	NA	1.18	NA	0.23
1.6	PC-A	3.0	12	1	1:1	0	NA	1.19	NA	0.30
1.7	PC-A	3.0	6	1	2:1	0	NA	1.25	NA	0.35
1.8	PC-A	3.0	12	1	2:1	0	NA	0.98	NA	0.40
1.1	FC-S-0	0.7	0	1	1:1	0	1.03	1.17	0.73	0.46
1.2	FC-S-0	0.7	6	1	1:1	0	0.87	1.12	0.90	0.50
1.3	FC-S-0	0.7	12	1	1:1	0	1.10	1.13	0.90	0.62
1.4	PC-A	0.7	0	1	1:1	0	1.07	1.19	0.67	0.27
1.5	PC-A	0.7	6	1	1:1	0	1.28	1.30	0.63	0.25
1.6	PC-A	0.7	12	1	1:1	0	1.10	1.14	0.56	0.33
Multiple barrel experiments										
2.1	FC-S-0	0.7	6	1	1:1	0	0.87	1.12	0.90	0.50
2.2	FC-S-30	0.7	6	1	1:1	0	1.07	1.16	0.71	0.26
2.3	FC-D-0	0.7	6	2	1:1	0	0.84	1.07	0.71	0.52
2.4	FC-D-0-E	0.7	6	2	1:1	0	0.86	1.09	0.32	0.53
2.5	FC-D-30	0.7	6	2	1:1	0	0.76	1.03	0.74	0.34
2.6	FC-D-30-E	0.7	6	2	1:1	0	0.90	1.09	0.42	0.31
2.7	FC-T-0	0.7	6	3	1:1	0	1.00	0.91	0.60	0.54
2.8	FC-T-0-E	0.7	6	3	1:1	0	0.90	0.93	0.80	0.58

1 inch = 2.54 cm

124

Table 19. Outlet control design coefficients, all experiments—*Continued*.

Inlet	Model	Slope (percent)	Fillet (inches)	Barrels	Span: Rise	Skew (degrees)	Outlet loss K_o unsubmerged	Outlet loss K_o submerged	Entrance loss K_e unsubmerged	Entrance loss K_e submerged
2.9	FC-T-30	0.7	6	3	1:1	0	1.00	1.05	0.48	0.31
2.10	FC-T-30-E	0.7	6	3	1:1	0	1.09	1.26	0.41	0.32
2.11	FC-Q-0	0.7	6	4	1:1	0	1.23	1.09	0.83	0.52
2.12	FC-Q-0-E	0.7	6	4	1:1	0	1.08	1.16	0.87	0.50
2.13	FC-Q-30	0.7	6	4	1:1	0	0.86	1.08	0.38	0.32
2.14	FC-Q-30-E	0.7	6	4	1:1	0	0.89	1.05	0.38	0.34
2.15	PC-A	0.7	12	1	1:1	0	1.10	1.14	0.56	0.33
2.16	PC-B	0.7	12	2	1:1	0	0.79	1.00	0.96	0.49
2.17	PC-B-E	0.7	12	2	1:1	0	0.96	1.08	0.75	0.56
2.18	PC-C	0.7	12	3	1:1	0	1.01	1.25	0.94	0.54
2.19	PC-C-E	0.7	12	3	1:1	0	1.21	1.14	0.96	0.51
2.20	PC-D	0.7	12	4	1:1	0	1.05	0.98	0.91	0.59
2.21	PC-D-E	0.7	12	4	1:1	0	1.07	1.05	0.93	0.58
Span-to-rise experiments										
3.1	FC-S-0	0.7	0	1	1:1	0	1.03	1.17	0.73	0.46
3.2	FC-S-0	0.7	0	1	2:1	0	0.87	1.02	0.48	0.40
3.3	FC-S-0	0.7	0	1	3:1	0	1.20	1.45	0.66	0.32
3.4	FC-S-0	0.7	0	1	4:1	0	1.07	1.16	0.62	0.40
3.5	FC-S-30	0.7	0	1	1:1	0	0.75	1.12	0.39	0.27
3.6	FC-S-30	0.7	0	1	2:1	0	0.97	1.07	0.39	0.22

1 inch = 2.54 cm

Table 19. Outlet-control design coefficients, all experiments—*Continued*.

Inlet	Model	Slope (percent)	Fillet (inches)	Barrels	Span: Rise	Skew (degrees)	Outlet loss K_o unsubmerged	Outlet loss K_o submerged	Entrance loss K_e unsubmerged	Entrance loss K_e submerged
3.7	FC-S-30	0.7	0	1	3:1	0	1.49	1.27	0.48	0.19
3.8	FC-S-30	0.7	0	1	4:1	0	0.87	1.14	0.53	0.18
3.9	PC-A	0.7	0	1	1:1	0	1.07	1.19	0.67	0.27
3.10	PC-A	0.7	0	1	2:1	0	0.92	1.09	0.42	0.34
3.11	PC-A	0.7	0	1	3:1	0	1.76	1.52	0.80	0.29
3.12	PC-A	0.7	0	1	4:1	0	0.95	1.15	0.69	0.26
Skewed headwall experiments										
4.1	FC-T-0	0.7	0	1	1:1	0	1.31	1.32	0.84	0.39
4.2	FC-T-30	0.7	0	1	1:1	0	0.86	0.98	0.47	0.35
4.3	FC-T-30	0.7	0	1	1:1	15	0.88	1.13	0.86	0.47
4.4	FC-T-30	0.7	0	1	1:1	30	0.94	1.00	0.43	0.36
4.5	FC-T-30	0.7	0	1	1:1	45	0.95	0.97	0.85	0.44
4.6	FC-S-30	0.7	0	1	1:1	30	0.84	1.03	0.90	0.46

1 inch = 2.54 cm

126

APPENDIX D. EXAMPLE PROBLEM

The 25-year and 100-year floods at a 34.97-square-kilometer (km^2) (13.5-square-mile (mi^2)) design site in South Dakota have peak flows of 21.6 m^3/s (773 ft^3/s) (Q$_{25}$) and 44.9 m^3/s (1602 ft^3/s) (Q$_{100}$).

REQUIREMENT: Design and compare the headwater elevations for the Q$_{25}$ and Q$_{100}$ peak flows using a twin 2.7- by 2.4-m (9- by 8-ft) cast-in-place (field cast) culvert and a twin 2.7- by 2.4-m (9- by 8-ft) precast box culvert.

The low roadway grade has an elevation of 27.51 m (90.20 ft).

Given:

Elevation of inlet invert:	24.04 m (78.81 ft)
Elevation of outlet invert:	24.03 m (78.79 ft)
Culvert length:	25.62 m (84 ft)
Stream bed slope:	0.02 percent

The downstream cross section ground point coordinates are given in table 20.

Table 20. Example problem, downstream cross section ground point coordinates.

X (ft)	Y-Elevation (ft)	[Head?]
64	86.0	
130	84.0	
152	83.5	
197	83.0	
245	82.5	
277	82.0	
293	81.0	Edge of channel
300	78.7	
305	81.0	
329	82.0	
406	82.5	Edge of channel
470	83.0	
500	86.0	

1 ft = 0.305 m

The tailwater rating information is given in table 21.

Table 21. Example problem, tailwater rating information.

Flow (ft³/s)	Tailwater elevation (ft)
68.7	82.77
222.0	83.49
375.4	83.93
528.7	84.30
682.0	84.61
773.0	84.78
988.7	85.15
1142.0	85.38
1295.3	85.61
1448.7	85.82
1602.0	86.00

1 ft = 0.305 m; 1 ft³/s = 0.028 m³/s

STEP 1: Plot the downstream discharge rating curve and flow area curves based on ground point coordinates.

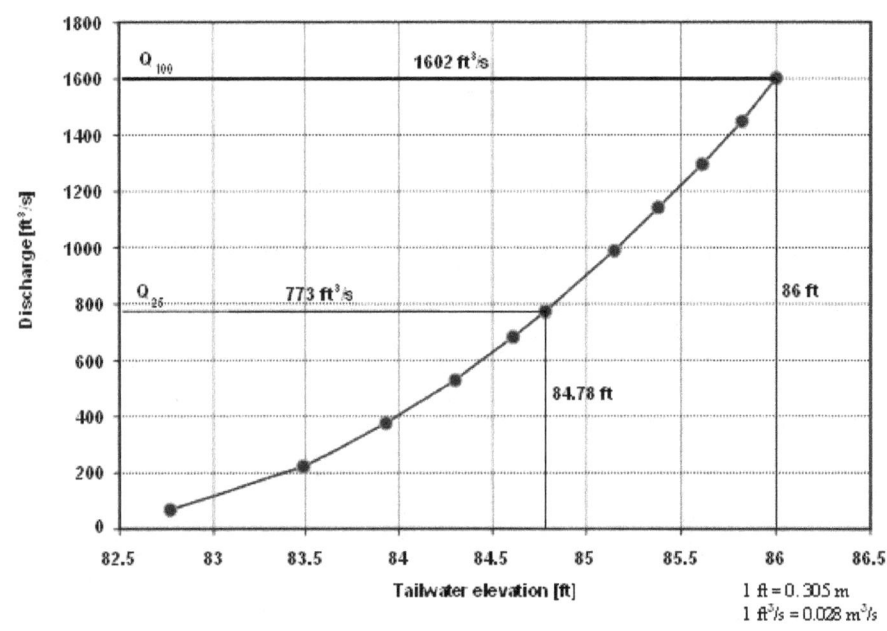

Figure 133. Graph. Discharge, tailwater variation.

128

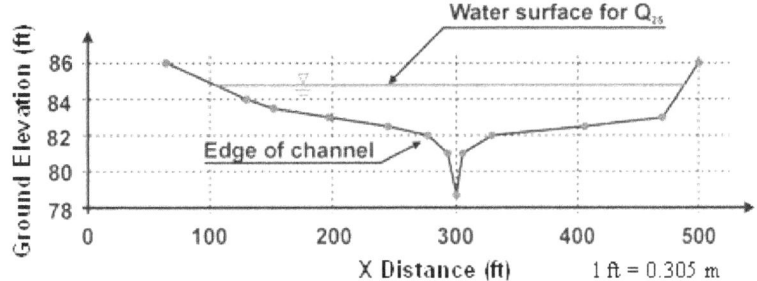

Figure 134. Graph. Downstream cross section.

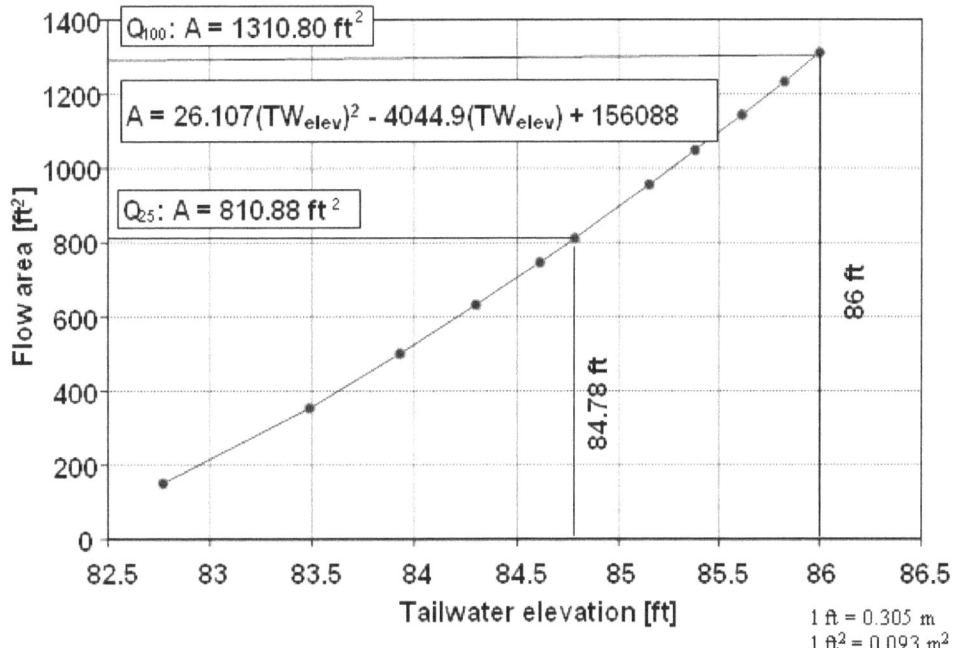

Figure 135. Graph. Cross section area versus tailwater elevation.

Based on a regression of the area versus the tailwater elevation curve in figure 135, the downstream flow area for tailwater elevation is given by the equation in figure 136.

$$A = 26.107(TW_{elve})^2 - 4044.9(TW_{elve}) + 156088$$

Figure 136. Equation. Downstream flow area for tailwater elevation.

Assuming the upstream section is a vertical shift of the downstream section according to the 0.02 percent channel slope, the flow area under the headwater elevation can be computed by the equation in figure 137.

129

$$A = 26.107(HW_{elve})^2 - 4045.777(HW_{elve}) + 156155.95$$

Figure 137. Equation. Flow area under headwater elevation.

<u>STEP 2</u>: Compute downstream channel velocity, V_{TW}.

$$V_{TW} = Q / A = 773 / 810.88 = 0.95 \; ft / s \; for \; Q_{25}$$
$$(0.95 \; ft / s = 0.29 \; m / s)$$

Figure 138. Equation. Downstream channel velocity for Q_{25}.

$$V_{TW} = Q / A = 1602 / 1310.80 = 1.22 \; ft / s \; for \; Q_{100}$$
$$(1.22 \; ft / s = 0.37 \; m / s)$$

Figure 139. Equation. Downstream channel velocity for Q_{100}.

<u>STEP 3</u>: Compute critical depth using the equation in figure 140.

$$d_c = \frac{(Q^2 B / g)^{1/3} + NBa^2}{B}$$

Figure 140. Equation. Critical depth, below top corner fillets.

Where:

B	is total culvert width; NB times span of each barrel.
d_c	is flow depth measured from the invert.
a	is the corner fillet height; 0.153 m (0.5 ft) for the FC culvert, and 0.305 m (1 ft) for the PC culvert.
NB	is the number of barrels.

The equation in figure 140 applies if the critical depth is below the top corner fillets. If the critical depth does partially submerge the top fillets, the relationship becomes the equation in figure 141.

$$d_c = \frac{(Q^2 (B - 2NBa_t) / g)^{1/3} + NB(a^2 + a_t^2)}{B}$$

Figure 141. Equation. Critical depth, partially submerged top corner fillets.

130

Where:

a_t is the submergence of top corner fillets; $d_c-(D-a)$.

Solving this equation would be a trial and error procedure, but it can be solved using the goal seek tool of Microsoft® Excel.

Table 22. Example problem, step 3 solutions.

Culvert (ft by ft)	a (inches)	d_c for Q_{25} (ft)	d_c for Q_{100} (ft)	Critical depth elevation at outlet for Q_{25} (ft)	Critical depth elevation at outlet for Q_{100} (ft)
FC-D-30 9 by 8	6	3.88	6.29	82.67	85.09
FC-D-0 9 by 8	6	3.88	6.29	82.67	85.09
PC-B 9 by 8	12	3.96	6.37	82.75	85.16
PC-B 9 by 8	0	3.86	6.26	82.65	85.05

1 inch = 2.54 cm; 1 ft = 0.305 m

STEP 4: Determine normal depths in the culvert from Manning's equation (figure 142).

$$Q = \frac{1.49}{n} A R_h^{2/3} S_o^{1/2}$$

Figure 142. Equation. Normal culvert depth.

If the normal depth is below the top corner fillet, the flow area, A, and the hydraulic radius, R_h, can be computed from the equations in figure 143.

$$A = NB(d_n span - a^2)$$
$$R_h = A/(NB(span + 2d_n - 1.17a))$$

Figure 143. Equations. Flow area and hydraulic radius, depth below top corner fillet.

If the normal depth partially submerges the top corner fillets, the flow area, A, and the hydraulic radius, R_h, can be computed from the equations in figure 144.

131

$$A = NB(d_n span - (a^2 + a_t^2))$$

$$R_h = A/(NB(span + 2d_n - 1.17a + 0.848a_t^2))$$

**Figure 144. Equations. Flow area and hydraulic radius,
top fillets partially submerged.**

Where:

a_t is the submergence of top corner fillets; $d_c - (D-a)$.

If the normal depth exceeds the rise, D, of the culvert, set a_t equal to a and compute the normal depth that would occur if the culvert did not have a crown.

The normal depth can be determined by trial and error or by using the goal seek tool from Excel.

Table 23. Example problem, step 4 solutions.

Culvert (ft by ft)	a (inches)	d_n for Q_{25} (ft)	d_n for Q_{100} (ft)
FC-D-30 9 by 8	6	10.49	19.61
FC-D-0 9 by 8	6	10.49	19.61
PC-B 9 by 8	12	10.69	19.82
PC-B 9 by 8	0	10.44	19.57

1 inch = 2.54 cm; 1 ft = 0.305 m

STEP 5: Determine initial depth d_o at barrel exit to start backwater calculation.

The normal depths are greater than critical depths; therefore, the culverts will be outlet control whether or not the barrels flow full at the inlet. The tailwater elevations are greater than the critical depth elevations and are below the crown elevations at the outlet; therefore, the depth at the culvert outlet will be between the critical depth and the culvert crown and can be computed from the equation in figure 145.

$$d_o + \frac{V_o^2}{2g} = TW \; Elevation + \frac{V_{TW}^2}{2g} - Invert \; Elevation + K_o\left(\frac{V_o^2}{2g} - \frac{V_{TW}^2}{2g}\right)$$

Figure 145. Equation. Initial depth.

Assume $K_o = 1.0$ (see HDS-5, p. 35).

132

The outlet is unsubmerged, and the footnote below table 6 of the research report warns that the unsubmerged K_o values are unreliable. Nevertheless, the unsubmerged value for a twin box culvert happens to be the traditional value that is recommended for the outlet loss. The exit coefficients derived for this study neglect the tailwater velocity head.

$$for\ K_o = 1.0 \rightarrow d_o = TW\ Elevation - Invert\ Elevation$$

Figure 146. Equation. For K_o equals 1.0.

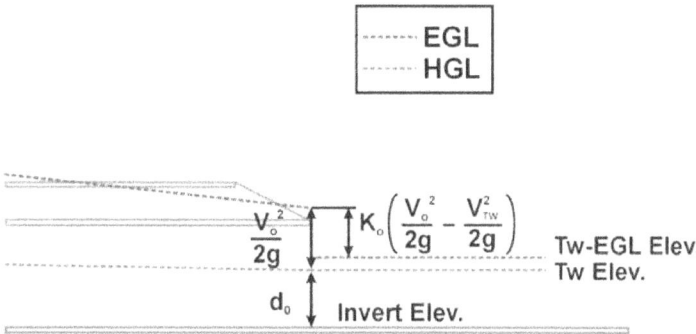

Figure 147. Diagram. Definition sketch for exit loss.

$$d_o = TW\ Elevation - Invert\ Elevation\ at\ outlet = TW\ Elevation - 78.79$$

Figure 148. Equation. Initial depth, ignoring tailwater velocity head.

Table 24. Example problem, step 5 solutions.

Culvert (ft by ft)	TW HGL elevation for Q_{25} (ft)	d_o for Q_{25} (ft)	EGL elevation at culvert outlet for Q_{25} (ft)	TW HGL elevation for Q_{100} (ft)	d_o for Q_{100} (ft)	EGL elevation at culvert outlet for Q_{100} (ft)
FC-D-30 9 by 8	84.78	5.99	85.58	86.00	7.21	88.38
PC-B 9 by 8	84.78	5.99	85.61	86.00	7.21	88.46

1 ft = 0.305 m

133

STEP 6: Use standard step backwater calculations to determine the EGL in the culvert for free surface flow.

This step can be done fairly easily on a spreadsheet by increasing the depth by increments between d_o at the outlet and the full culvert depth, D. Compute the step length, ΔL, from the equation in figure 153. Figures 149–152 contain equations for preliminary calculations for the equation in figure 153.

$$\Delta d = [(D-a)-d_o]/1000; \textit{ take } 1000 \textit{ steps in the spreadsheet}$$
$$d_i = d_{i-1} + \Delta d$$
$$A_i = NB(d_i span - a^2)$$
$$R_{hi} = A_i /(NB(span + 2d_i - 1.17a))$$

Figure 149. Equations. For d less than (D−a).

$$\Delta d = a/20; \textit{ take } 19 \textit{ steps in the spreadsheet}$$
$$d_i = d_{i-1} + \Delta d$$
$$a_{ti} = d_i - (D-a)$$
$$A_i = NB(d_i span - (a^2 + a_{ti}^2))$$
$$R_{hi} = A_i /(NB(span + 2d_i - 1.17a + 0.848a_{ti}))$$

Figure 150. Equations. For d less than D but greater than (D−a).

$$d_i = D$$
$$A_i = NB(D span - 2a^2)$$
$$R_{hi} = A_i /(NB(2 span + 2D - 2.34a))$$

Figure 151. Equations. For d equal to D (the last iteration).

Where:

subscript "i"   is a line in the step-backward computation.
a_t is the partially submerged top fillet.
d   is the flow depth in the barrel.
D   is the rise of the culvert.

The friction slope, S_F, for any step is computed from Manning's equation.

$$S_F = \left(\frac{nQ}{1.5 A_m R_{hm}^{2/3}} \right)^2$$

Figure 152. Equation. Friction slope.

134

Where:

A_m   is $(A_i + A_{i-1})/2$.
R_hm   is $(R_{hi} + R_{hi-1})/2$.
n   is the Manning roughness coefficient: 0.008 m$^{-1/3}$ (0.012 ft$^{-1/3}$).
Q   is discharge.

The results of the equations in figures 149–152 enable the calculation of the step length, ΔL. The step length may also be viewed as calculated from the energy balance.

$$\Delta L = \frac{\left(d_i + \dfrac{V_i^2}{2g}\right) - \left(d_{i+1} + \dfrac{V_{i+1}^2}{2g}\right)}{S_b - S_F}$$

Figure 153. Equation. Step length.

Where:

V   is flow velocity in barrel.
S_b   is barrel slope.

After each step length calculation, the ΔL's are summed to give L, which is compared with the length of the culvert (25.62 m (84 ft)) to determine when computations are complete. Since the corner fillets affect the backwater computations, the calculations should be done in two stages.

If the culvert fills to the crown before the computations reach the entrance, the EGL at the entrance, or upstream end, is given by the equation in figure 154.

$$EGL_{US} = D + \frac{V_{FULL}^2}{2g} + \left(\frac{29n^2 L_{FULL}}{R_{hFULL}^{4/3}}\right)\frac{V_{FULL}^2}{2g}$$

Figure 154. Equation. EGL at upstream culvert end (the entrance).

Where:

V_FULL   is Q/A_{FULL}.
A_FULL   is $NB((span)D - a^2)$.
R_hFULL   is $A_{FULL}/(NB(span + 2D - 2.343a))$.
L_FULL   is the length of a culvert flowing full.

STEP 7: Determine entrance loss.

HGL is greater then critical depth throughout the barrel, indicating outlet control. Use table 11 (in chapter 7) to find entrance loss coefficients. The sketches in figure 155 are from figure 93 in chapter 7.

Inlet	Sketch	K_e
a. Sketch 2 30°-flared wingwalls; top edge beveled at 45°; 2 barrels (FC-D-30)	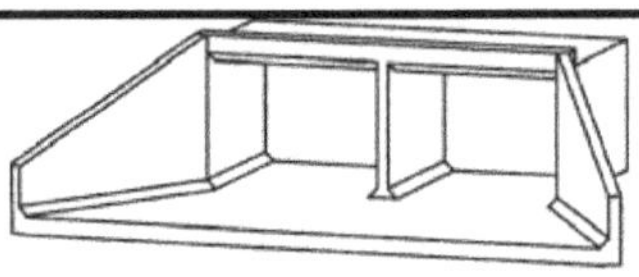	**0.32**
b. Sketch 8 0°-flared wingwalls (extended sides); top edge beveled at 45°; 2 barrels (FC-D-0)	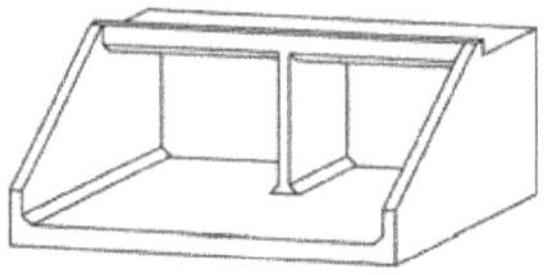	**0.52**
c. Sketch 12 0°-flared wingwalls (extended sides); crown rounded at 8-in. radius; 12-in. corner fillets; 2 barrels (PC-B)	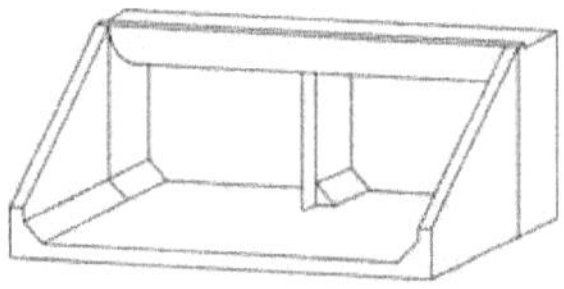	**0.54**

1 inch = 2.54 cm

Figure 155. Sketches. Entrance loss coefficients (K_e) of culverts in example problem.

$$h_{Le} = K_e \frac{V_{US}^2}{2g}$$

Figure 156. Equation. Entrance loss.

Where:

V_{US} is velocity at the upstream end of the culvert from the step backwater or the full flow computations.

STEP 8: Compute headwater elevation.

$$HW_{EGL} = EGL \; elevation_{US} + h_{Le}$$

Figure 157. Equation. Headwater energy grade line.

136

Where:

EGL elevation$_{US}$ is the EGL elevation at the upstream end of the culvert from the backwater calculations.

The HW$_{EGL}$'s are energy grade line elevations and include the velocity head, which is usually negligible in the headwater pool. To get the actual water surface elevations, HW$_{HGL}$ in the headwater pool, use the relationships in figure 158.

$$HW_{EGL} = HW_{HGL} + \frac{V_{HW}^2}{2g}, \quad where \ \frac{V_{HW}^2}{2g} = Q/A$$

Figure 158. Equation. Headwater hydraulic grade line.

Where:

A   is computed from the equation in figure 137.

The water surface elevation, HW$_{elevation}$, in the headwater pool can be determined by trial and error or by using the goal seek tool of Excel.

Tables 25 and 26 summarize the results of the step backwater computations, the headwater EGL computations, and the headwater HGL computations.

Table 25. Example problem, step backwater and entrance loss results for Q$_{25}$.

Culvert	Span, Rise	Fillet Size (inches)	HGL Elevation$_{US}$ (ft)	EGL Elevation$_{US}$ (ft)	K$_e$	HW$_{EGL}$ (ft)	Area in HW Pool (ft^2)	Velocity in HW Pool (ft/s)	Water Surface Elevation HW$_{elevation}$ (ft)
FC-D-30	9, 8	6	84.877	85.662	0.32	85.913	1265	0.65	85.907
FC-D-0	9, 8	6	84.877	85.662	0.52	86.071	1335	0.58	86.065
PC-B	9, 8	12	84.879	85.686	0.54	86.121	1358	0.57	86.116
PC-B	9, 8	0	84.879	85.656	0.54	86.076	1228	0.63	85.823

1 inch = 2.54 cm; 1 ft = 0.305 m

137

Table 26. Example problem, step backwater and entrance loss results for Q_{100}.

Culvert	Span, Rise	Fillet Size (inches)	HGL Elevation$_{US}$ (ft)	EGL Elevation$_{US}$ (ft)	K_e	HW$_{EGL}$ (ft)	Area in HW Pool (ft^2)	Velocity in HW Pool (ft/s)	Water Surface Elevation HW$_{elevation}$ (ft)
FC-D-30	9, 8	6	86.458	88.578	0.32	89.256	3028	0.53	89.251
FC-D-0	9, 8	6	86.458	88.578	0.52	89.680	3294	0.48	89.676
PC-B	9, 8	12	86.471	88.657	0.54	89.837	3394	0.47	89.834
PC-B	9, 8	0	86.463	88.563	0.54	89.697	3305	0.48	89.693

1 inch = 2.54 cm; 1 ft = 0.305 m

For the four culvert configurations, results of the step backwater and entrance loss computations for the Q_{100} discharge only are illustrated in figures 159 through 162.

138

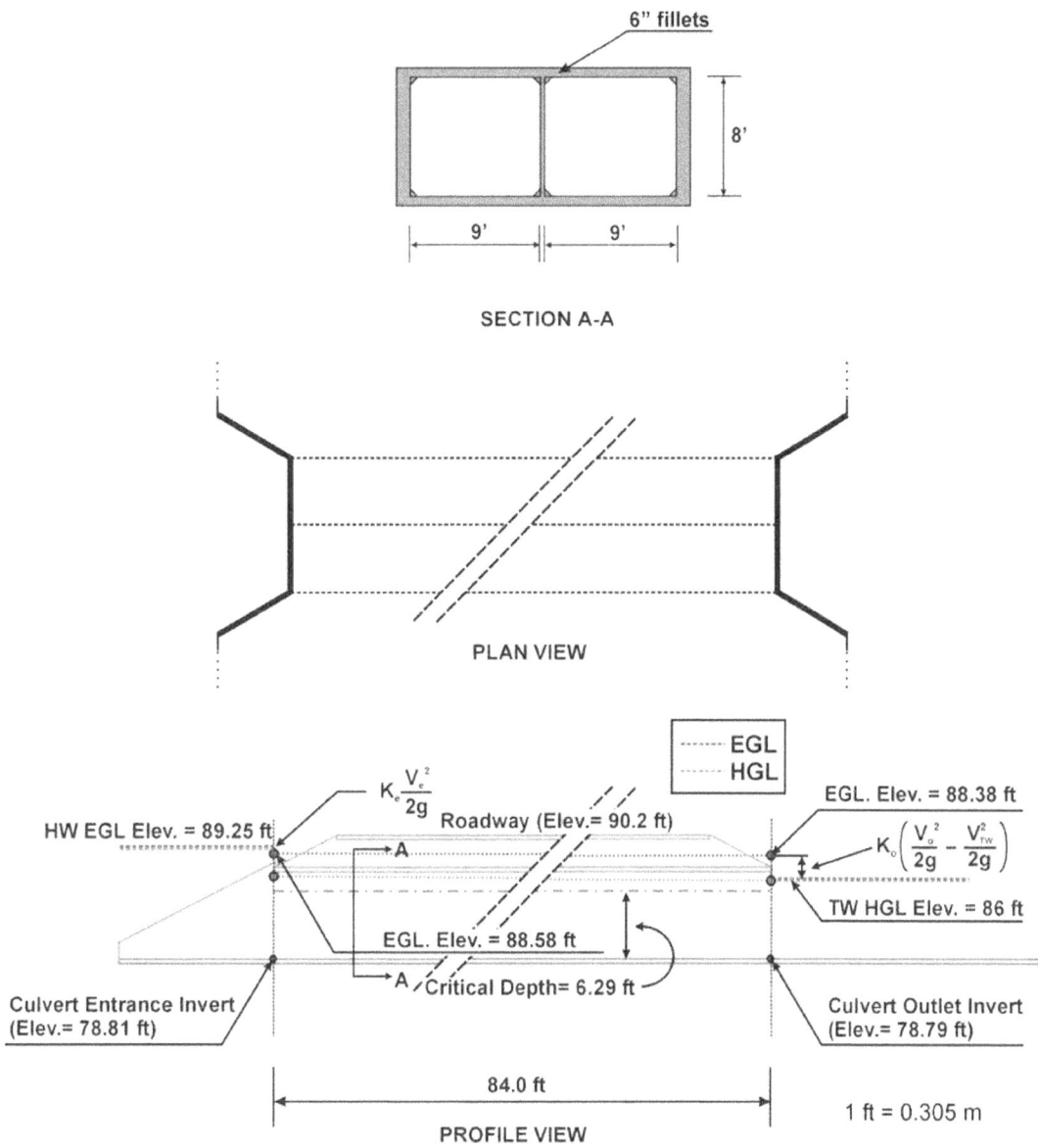

Figure 159. Diagram. FC-D-30 model, Q$_{100}$ elevations.

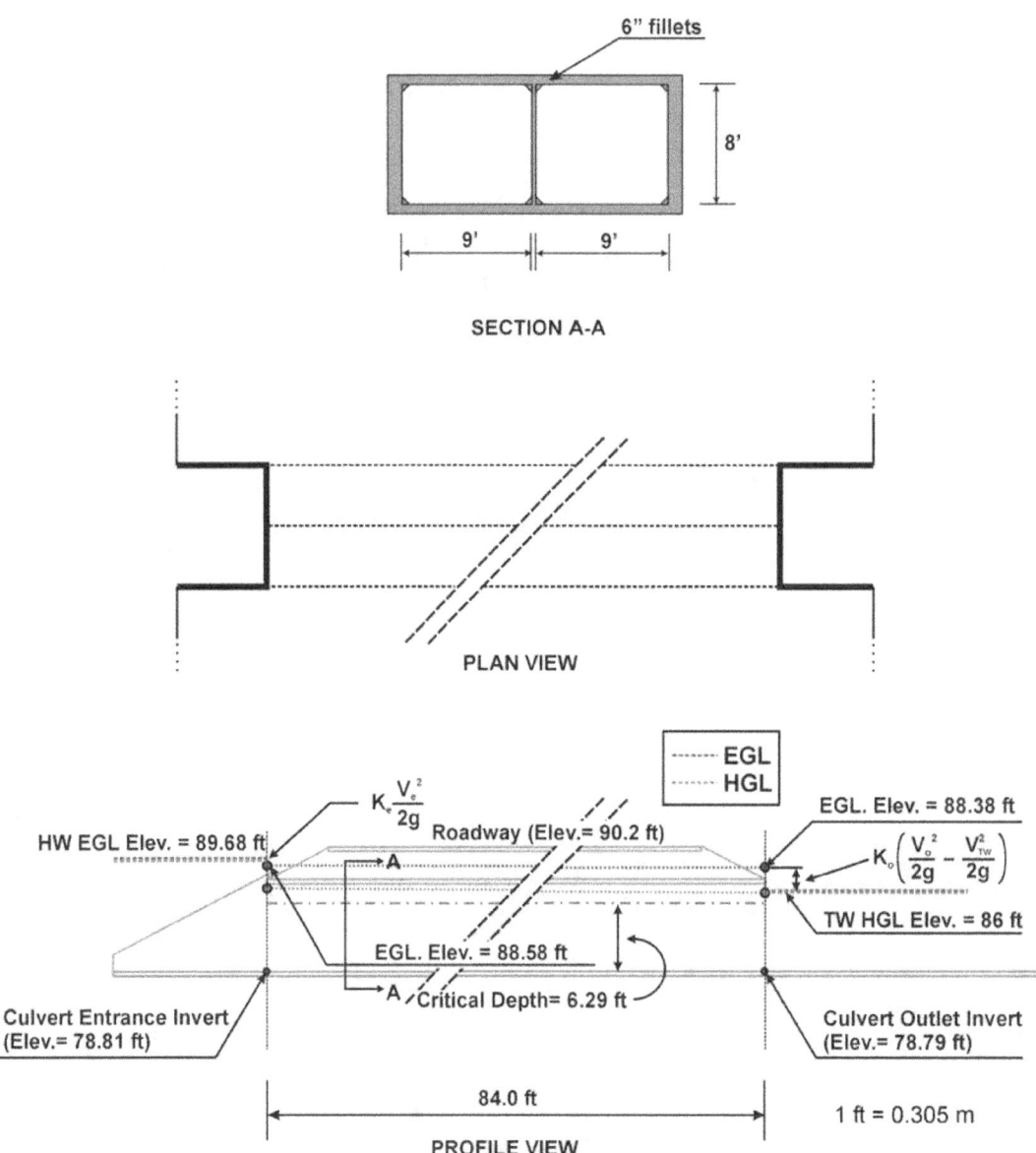

Figure 160. Diagram. FC-D-0 model, Q_{100} elevations.

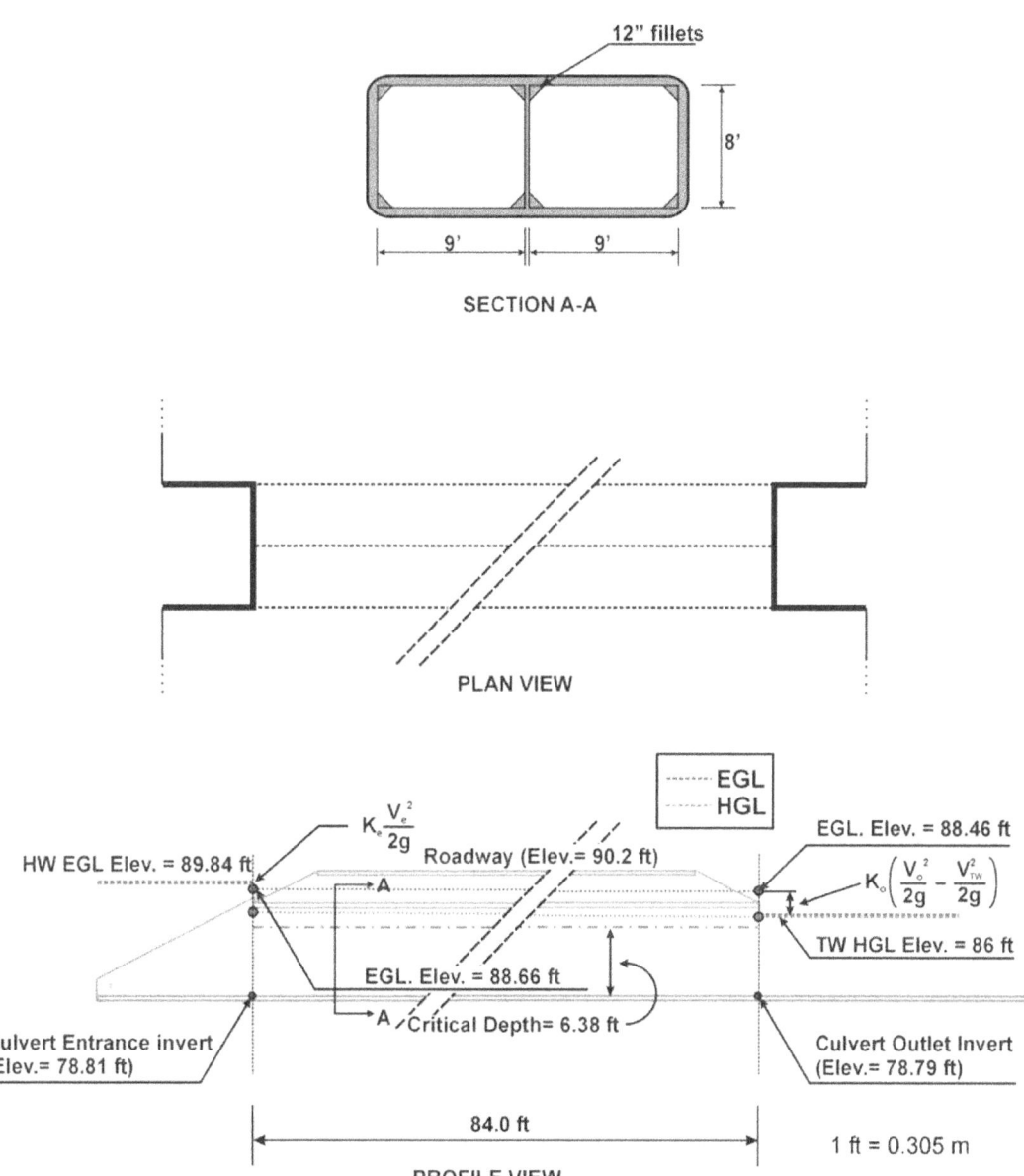

Figure 161. Diagram. PC-B model, 30.48-cm (12-inch) corner fillets, Q_{100} elevations.

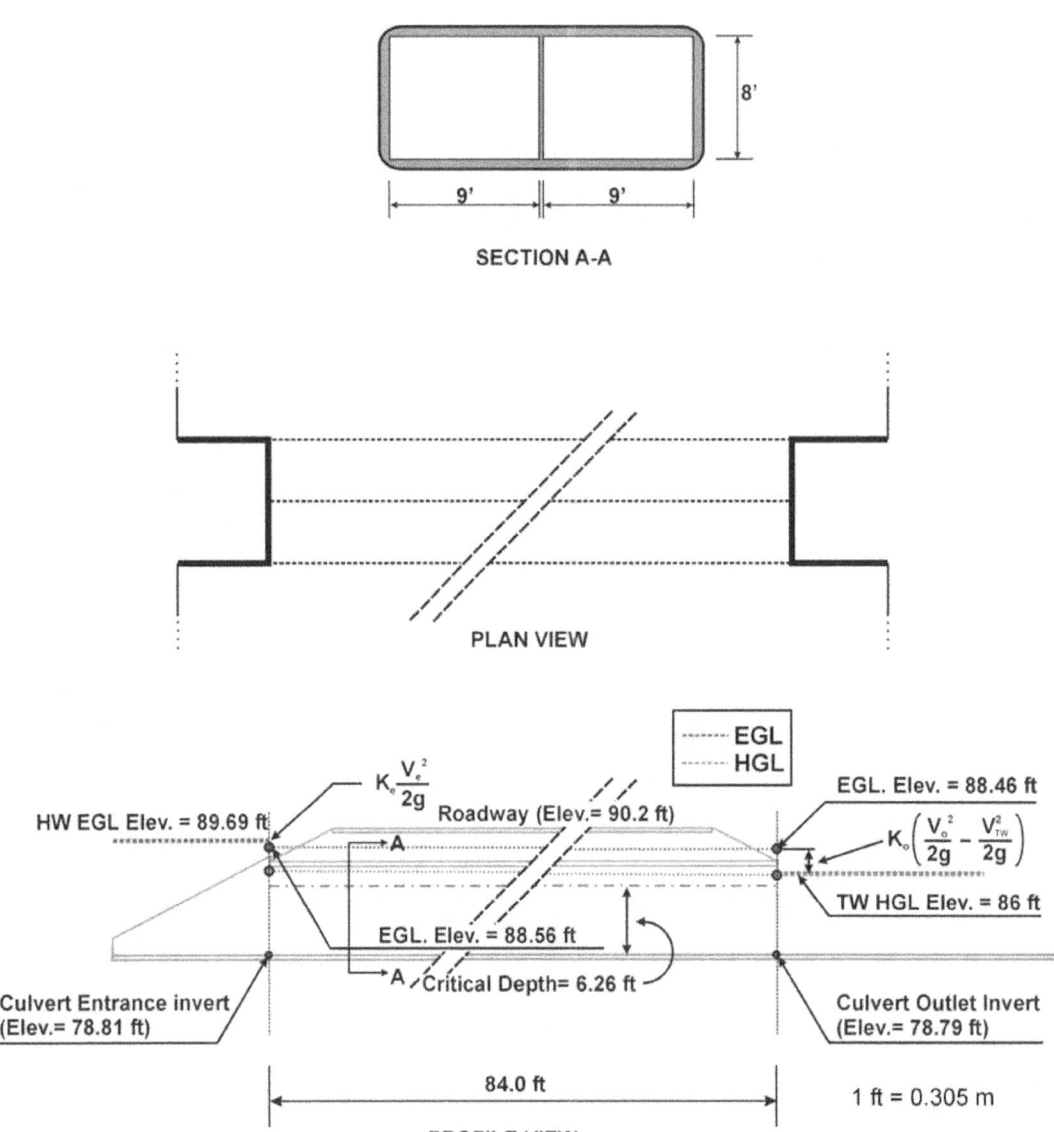

Figure 162. Diagram. PC-B model, no corner fillets, Q₁₀₀ elevations.

SUMMARY

This example is included to illustrate how to apply the entrance loss coefficients from the laboratory results. SDDOT provided site data and culvert options. The example was divided into eight basic steps for spreadsheet computations because no design program accounted for the corner fillets, which were a consideration in the laboratory study.

The first step was to plot the channel cross section and derive expressions for channel area versus water surface elevations. The tailwater channel velocities and EGL elevations were then computed from the downstream rating data that were provided. The critical depth and normal depth in the culvert were computed to determine if inlet control was a possibility. The normal depth computation is a tedious trial and error process, especially when corner fillets are included in the computation, but the goal seek tool from Excel makes the task easier. Because normal depths were greater than critical depths, inlet control was eliminated as a possibility. The brink depth at the culvert outlet was computed from the equations in figure 163.

$$d_o = EGL_o - \frac{V_o^2}{2g}$$

$$EGL_o = EGL_{TW} + h_{Lo}$$

$$h_{Lo} = K_o \left(\frac{V_o^2}{2g} - \frac{V_{TW}^2}{2g} \right)$$

$$V_o = \frac{Q}{d_o \, (total \; culvert \; width)}$$

Figure 163. Equations. Brink depth at culvert outlet.

Where:

K_o is the outlet loss coefficient, assumed to be 1.0.

Since the brink depth was below the crown of the culvert, a spreadsheet was developed for step backwater computations through the culvert by increasing the depth by increments and computing the corresponding step length. Spreadsheets were analyzed to determine where either the cumulative step lengths equaled the culvert length or the culvert flowed full. None of the culverts flowed full for either Q_{25} or Q_{100} before the cumulative step lengths equaled the culvert length of 25.62 m (84 ft); thus free surface flow occurred in each case. From the step backwater computations, the velocity and energy grade line elevation at the upstream end were used to compute the headwater energy grade line elevation from the equation in figure 164.

143

$$HW_{EGL} = EGL_{US} + K_e \frac{V_{US}^2}{2g}$$

Figure 164. Equation. Headwater EGL.

Where:

EGL_{US} and V_{US} are energy grade line elevation and velocity at the upstream end of the culvert from the step backwater computations.

K_e is the entrance loss coefficient from the laboratory results: 0.32 for the field cast inlet with 30-degree-flared wingwalls; 0.52 for the field cast inlet with 0-degree-flared wingwalls; and 0.54 for the precast inlet.

Finally, the hydraulic grade line elevation (water surface elevation) was computed by subtracting the velocity head in the headwater pool from the energy grade line elevation. Because of the irregular channel geometry, this was also a trial and error computation and was accomplished with the Excel goal seek tool. There was very little difference between the energy grade line and water surface elevations in the headwater pool.

The net area (with the areas of fillets removed) was used for the step backwater computations and for velocity computations. The research showed that the coefficients were not affected by the fillet sizes tested as long as the net area was used in the computations. Figure 165 is a sketch showing the procedure for calculating the net area. The current version of the FHWA HY-8 program does not account for corner fillets.

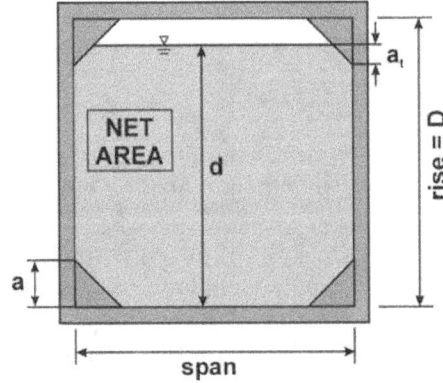

Figure 165. Diagram. Net area used for backwater computations.

To show the sensitivity of including or not including the corner fillets, the headwater elevation for the PC-B, 2.7- by 2.4-m (9- by 8-ft) culverts was computed with no fillets. The error derived by not accounting for the 30.48-cm (12-inch) fillets was more than 0.0305 m (0.10 ft) for the Q_{100} discharge. These errors would increase as the size of the culvert decreases and would decrease as the size of the culvert increases. A design program such as HY-8 could certainly account for the corner fillets, but significant additional computer coding would be required.

At the Q_{100} discharge, the HW elevations for the precast culverts were approximately 0.183 m (0.6 ft) higher than the elevations for the field cast culverts with the 30-degree-flared wingwalls.

REFERENCES

1. Normann, J. M., Houghtalen, R. J., and Johnston, W. J. *Hydraulic Design of Highway Culverts*, Federal Highway Administration Hydraulic Design Series No. 5 (HDS-5), Report Number FHWA-IP-85-15, McLean, VA, September 1985.

2. Federal Highway Administration, HY-8 , HDS-5 Appendix D Chart Calculator, www.fhwa.dot.gov/engineering/hydraulics/software.

3. Graziano, F., Stein, S., Umbrell, Ed., and Martin, B. *South Dakota Culvert Inlet Design Coefficients*, Federal Highway Administration, Report Number FHWA-RD-01-076, McLean, VA, June 2001.

4. French, J. L. *Hydraulics of Short Pipes, Hydraulic Characteristics of Commonly Used Pipe Entrances, First Progress Report*, National Bureau of Standards, NBS Report Number 4444, Washington, DC, December 1955.

5. French, J. L. *Hydraulics of Culverts, Second Progress Report, Pressure and Resistance Characteristics of a Model Pipe Culvert*, National Bureau of Standards, NBS Report Number 4911, Washington, DC, October 1956.

6. French, J. L. *Hydraulics of Culverts, Third Progress Report, Effects of Approach Channel Characteristics on Model Pipe Culvert Operation*, National Bureau of Standards, NBS Report Number 5306, Washington, DC, June 1957.

7. French, J. L. *Hydraulics of Improved Inlet Structures for Pipe Culverts, Fourth Progress Report, Pressure and Resistance Characteristics of a Model Pipe Culvert*, National Bureau of Standards, NBS Report Number 7178, Washington, DC, August 1961.

8. French, J. L. *Hydraulics of Culverts, Fifth Progress Report, Non-enlarged Box Culvert Inlets*, National Bureau of Standards, NBS Report Number 9327, Washington, DC, June 1966.

9. French, J. L. *Hydraulics of Culverts, Sixth Progress Report, Tapered Box Culvert Inlets*, National Bureau of Standards, NBS Report Number 9355, Washington, DC, June 1966.

10. French, J. L., and Bossy, H. G. *Hydraulics of Culverts, Seventh Progress Report, Tapered Box Inlets with Fall Concentration in the Inlet Structure*, National Bureau of Standards, NBS Report Number 9528, Washington, DC, July 1967.

11. Jones, J. S., Mistichelli, M. P., and Kilgore, R. T. *Long Span and Special Shape Culverts*, Federal Highway Administration, unpublished laboratory report, McLean, VA, June 1991.

12. Graziano, F., Stein, S., Umbrell, E., and Martin, B. *Hydraulics of Iowa DOT Slope-Tapered Pipe Culverts*, Federal Highway Administration, Report FHWA-RD-01-077, McLean, VA, June 2001.

13. GKY and Associates. *Compilation of Culvert Design Coefficients*, Federal Highway Administration, unpublished lab report, McLean, VA, October 1998.

14. McEnroe, B. M., and Johnson, L. M. *Hydraulics of Flared End Sections for Pipe Culverts*, Transportation Research Board, Transportation Research Record 1483, Washington, DC, 1995.

15. Umbrell, E., Young, K., Estes, C., Stein, S., and Pearson, D. R. *Hydraulics of Dale Boulevard Culverts: Performance Curve for a Prototype of Two Large Culverts in Series*, Federal Highway Administration, unpublished report, McLean, VA, July 2001.

16. Tullis, B. Personal communication, August 2004.

www.ingramcontent.com/pod-product-compliance
Lightning Source LLC
Chambersburg PA
CBHW080811180526
45168CB00006B/2400